EVOLUTION & CREATIONISM

A Very Short Guide

Second Edition

Warren D. Allmon

Paleontological Research Institution
Ithaca, New York
2009

ISBN 978-0-87710-484-1
Library of Congress catalog number 2009922136

Paleontological Research Institution Special Publication No. 35

1259 Trumansburg Road
Ithaca, New York 14850 U. S. A.
www.priweb.org

Cover design and layout by Paula Mikkelsen.

On the cover (counterclockwise from upper right): Fossil collectors on the cliffs at Red Hill, Pennsylvania; children at the Museum of the Earth examine fossils; the jungle at Sierra Madre, as a symbol for high levels of biodiversity (photograph by Perojevic via Wikimedia Commons); the U.S. Supreme Court; Birds of Paradise (from *Brehms Thierleben, Allgemeine Kunde des Thierreichs, 2nd ed.*, Bibliographisches Institut, Leipzig, Germany, 1882) as one of Darwin's favorite subjects; a fossil ammonite from the Paleontological Research Institution specimen collections.

CONTENTS

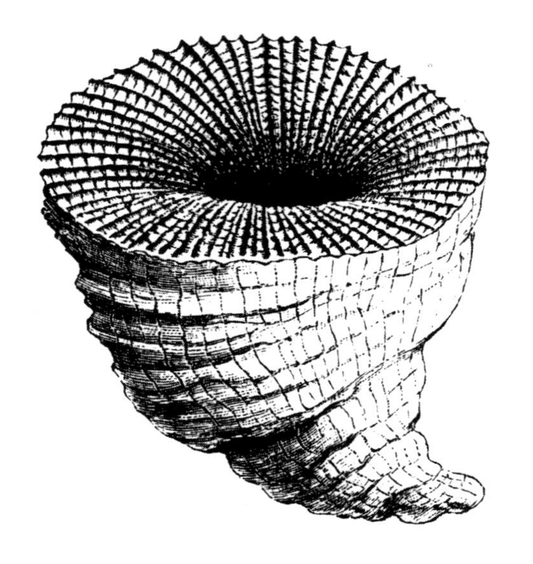

There is grandeur in this view of life, with its several powers, having been originally breathed into a few forms or into one; and that, whilst this planet has gone cycling on according to the fixed law of gravity, from so simple a beginning endless forms most beautiful and most wonderful have been, and are being, evolved.

- Charles Darwin, *On the Origin of Species* (1859), page 492

Yes, the world has been different ever since Darwin. But no less exciting, instructing, or uplifting; for if we cannot find purpose in nature, we will have to define it for ourselves. Darwin was not a moral dolt; he just didn't care to fob off upon nature all the deep prejudices of Western thought. Indeed, I suggest that the true Darwinian spirit might salvage our depleted world by denying a favorite theme of Western arrogance – that we are meant to have control and dominion over the earth and its life because we are the loftiest product of a preordained process. In any case, we must come to terms with Darwin. And to do this, we must understand both his beliefs and their implications.

- Stephen Jay Gould, *Ever Since Darwin* (1977), page 13

PREFACE & ACKNOWLEDGMENTS

There is no shortage of books about evolution and creationism. Why one more? This volume originated as a guide for volunteer docents at the Museum of the Earth in Ithaca, New York, designed to help them answer visitor questions about evolution and creationism. Despite the abundance of resources available about these subjects, we found to our surprise that there was no single source that could serve as a compact, concise, user-friendly handbook that addressed most of the issues that our docents needed to know about in a brief but authoritative way, and that also provided an easy entry into the huge existing literature. We have also found that teachers and the general public also need such a resource.

This book focuses on evolution and creationism because these two topics are (unfortunately) usually connected in the popular media, which promotes the idea that these two are equivalent and competing world views even among scientists. This is not true. As emphasized throughout this book, evolution is a highly supported scientific hypothesis, whereas creationism is a religious (and sometimes political) movement that has no scientific support.

This book is intended for readers at a variety of levels – from those with no more than a high school background in science, to teachers, college students, and professionals in other fields. In other words, it is intended for the general adult American public, all of whom (whether they realize it or not) really need to be familiar with the information summarized here. The central theme of this book is that an understanding of evolution can not only lead to a fuller and more satisfying understanding of living things, including ourselves, but is essential for making informed decisions about very real and immediate problems, from the environment to human health.

This second edition has much the same structure as the first (revised in 2006), but is otherwise almost completely rewritten, with much more discussion of natural selection and other causes of evolution and also of evolution's broader impacts, and a new chapter on teaching evolution.

Previous versions benefitted greatly from the assistance and comments of Emily Butler, Lenore Durkee, Amy McCune, Ed Picou, Rob Ross, Jennifer Tegan, and Will Provine. For the second edition, I am grateful to Richard Kissel, Amy McCune, Rob Ross, and Samantha Sands for comments on the manuscript, to Rob Bleiweiss, Kelly Cronin, Nancy Currier, and Richard Kissel for help with the illustrations, and to Kelly Cronin and especially Paula Mikkelsen for editorial assistance. Printing of this edition was made possible by the generous financial support of Mr. Arthur Kuckes.

1. INTRODUCTION

Organic evolution is the theory that all living things on Earth are connected by genealogy and have changed through time, or, as Charles Darwin more eloquently put it in 1859, that all living things are results of "descent with modification."[1] Among all scientific ideas and theories, evolution is unique. No other concept in science has produced so much controversy, debate, and emotion outside of scientific circles, and it continues to cause heated arguments today. This situation is remarkable for several reasons. Despite statements by its critics, and widespread misunderstanding by the general public to the contrary, there is *no* debate among credible scientists about whether evolution is true and a fully adequate explanation for what we observe about the history, order, and diversity of life on Earth, and there has been no serious debate about this issue among credible scientists for more than a century. This doesn't guarantee that evolution is actually true. It simply means that the vast majority of scientists accept that it is.

Evolution, indeed, is *the* central idea in modern biology. This means that it is the fundamental explanatory and theoretical underpinning that unites all aspects and subfields of biological science. Playing a similar role to atomic theory in physics and chemistry, or plate tectonics in geology, evolution is the basic modern scientific hypothesis for why living things are the way they are.

Furthermore, modern Western society beyond science has, to a remarkable degree, been shaped by this widespread scientific agreement. We largely take for granted, for example, that the world and everything in it have histories and change constantly. The scientific worldview of modern society is in part a product of the widespread acceptance of Darwin's theory of evolution by scientists. Evolution in general, and Darwin's particular explanation for it, have, furthermore, influenced almost all areas of the wider culture, from literature to economics to cartoons. In this sense, we all effectively live in Darwin's world. Yet, paradoxically, the majority of the general public does not know much about evolution, and much of what they think they know they do not accept.

Why does this matter? Because evolution is as well-supported a scientific idea as any other we routinely accept as true, such as that the Earth revolves around the Sun, that matter is made of atoms, that bacteria cause disease, or that the continents move. If evolution can be universally accepted by scientists and rejected by a majority of the general public, then so might any other highly verified scientific idea, threatening the rationality and basic scientific literacy that are crucial to the economic and social welfare of modern civilization. We live in a world increasingly dependent on science and technology. If we cannot understand how science and technology work, then we cannot make wise decisions on their use.

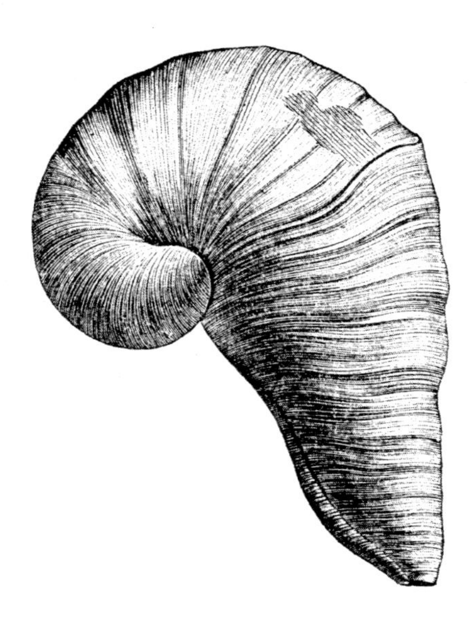

2. EVOLUTION & SCIENCE

Science is an approach to explaining and understanding the natural (= "material" or "physical") world. Its philosophical approach is frequently called **materialism** or **naturalism**. Science uses observations about the world and the rules of logic to test hypotheses that explain natural phenomena. **Hypotheses** are ideas about natural phenomena – they might or might not be true. Hypotheses can come from anywhere. What makes a hypothesis scientific is that it is testable. **Testing** hypotheses means making predictions from them, and then comparing these predictions to observations from the physical world. Hypotheses that pass such tests are accepted, but such acceptance is always **provisional**, that is, any accepted hypothesis can be overturned by sufficient credible contrary evidence. A **theory** in science is an idea or set of ideas and hypotheses that connects, explains, and is supported by a large number of observations. Although in common English, a "theory" can mean a mere guess or supposition, in science it is the basic unit of our understanding of reality. A theory in science is about as good as it gets.

A fundamental tool of all science is **extrapolation** (Figure 1). For example, if we drop a ball, we can measure how fast it falls. We can then use this result to apply to other falling objects. Scientific ideas and theories are essentially about applying what is known to what is unknown. Science deals not only with phenomena that can be observed directly. Indeed, the essence of science is to make an observation or experiment, and then use the results to predict what we will see in another instance. If we had to personally observe everything in science to be certain of it, no progress could ever really be made in our

Figure 1. *Extrapolation in science allows us to do experiments on small examples of problems (top left), and then apply the results to much larger experiments, the outcome of which cannot be known certainly in advance (top right). In exactly the same way, scientists can study fossils (bottom left) and make inferences about what they cannot observe directly, such as the soft tissues or behavior of the original animals (bottom right).* (Allosaurus *skull image courtesy of Bob Ainsworth; allosaur jaw reconstruction by Steveoc 86 via Wikimedia Commons.)*

understanding of the world. Most importantly, extrapolation works. It allows us to make successful predictions about nature.

What scientists provisionally accept as true is not decided by a vote of opinions. It is supposed to be decided by agreement of theory with observation about nature. In practice, however, what is accepted as true in a particular area by the overall scientific community is usually the majority view among those scientists who are specialists in that area. If most specialists who have devoted years to researching a topic accept a particular theory, then it is usually treated as true by other specialists who have not studied it in as great detail. If opinion is mixed, with no clear majority of specialists holding one view or another, the issue is usually labeled "poorly understood" or "controversial." If a view accepted by a majority of specialists is challenged by new information or interpretations, it will generally not be discarded – and the new view will not be widely accepted – until enough contrary examples have been put forward to convince a majority of specialists to change their minds. If this doesn't happen, the old theory stands. Thus, for example, in the early 1980s the theory that the dinosaurs became extinct because of collision of an asteroid with Earth was a minority view; most specialists favored Earth-based causes. Over the years, however, enough observations were collected to convince the majority of scientists who work on this problem that an asteroid or comet did strike the Earth at that time, and therefore today that is what is reported in textbooks as what "most scientists think," that is, what we – for the moment at least – think is true.

Science deals with the past as well as the present. Although we cannot directly observe or experiment on what went on before humans were around to witness it, we can observe the results of processes that occurred in the past, and compare these to the results of processes that we can observe in the present. Thus (although it is frequently more difficult) we *can* test hypotheses about the past just as we can in the present. The larger the number of independent tests that support a hypothesis, the more confident we become that the hypothesis is true. This is exactly the approach that police and forensic scientists use to solve crimes – they can figure out "whodunit" even though they did not witness the event by examining the clues left behind.

Science deals only with the physical or material world. It does not deal with the supernatural or with questions or issues for which no material or physical evidence exists. Science is about seeking material causes for material phenomena. This does not necessarily mean that the supernatural does not exist, or that science can answer all questions about everything. It simply means that the supernatural – those phenomena that cannot be examined in terms of tangible matter and energy – are not within the purview of science.

Scientific materialism has, by and large, served humans well. Although science is not a panacea for all human problems, and many aspects of the technology that have resulted from science are far from benign – such as environmental degradation or weapons of mass destruction – materialistic science has also created levels of health and comfort, and brought humans a level of control over our surroundings and lives, that would have been unthinkable just a few centuries ago.

Evolution is one of the best-supported ideas in science, that is, there is abundant evidence that it is true, so much that it would be irrational to reject it. Although all ideas in science are provisional, and can potentially be overturned by sufficient contrary evidence, evolution is as close to being a "fact" as any widely accepted scientific hypothesis.

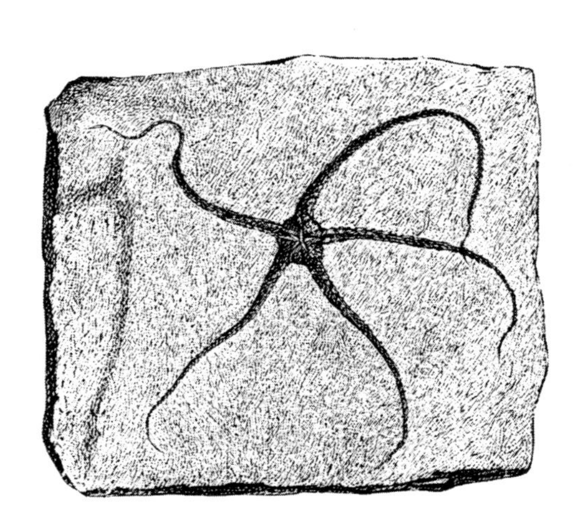

Summary - Evolution & Science:

- Science is a method for understanding the natural world. It uses observations about the world and the rules of logic to test hypotheses that explain natural phenomena. It has nothing to say about the supernatural.

- Hypotheses that pass such tests are accepted, but such acceptance is always provisional; that is, any accepted hypothesis can be overturned by sufficient credible contrary evidence.

- A theory in science is much more than a guess; it is an idea or set of ideas and hypotheses that connects, explains, and is supported by a large number of observations.

- A fundamental tool of all science is extrapolation. This is the use of inference to conclude what we did not or cannot actually see, and to build on the results of previous investigation.

- Science isn't "truth," but it is the best technique that humans have so far developed for discerning what the natural world is and how it works. Science is successful because it works.

- Evolution is as close to being a "fact" as any widely accepted scientific hypothesis.

ON

THE ORIGIN OF SPECIES

BY MEANS OF NATURAL SELECTION,

OR THE

PRESERVATION OF FAVOURED RACES IN THE STRUGGLE FOR LIFE.

BY CHARLES DARWIN, M.A.,

FELLOW OF THE ROYAL, GEOLOGICAL, LINNÆAN, ETC., SOCIETIES;
AUTHOR OF 'JOURNAL OF RESEARCHES DURING H. M. S. BEAGLE'S VOYAGE ROUND THE WORLD.'

Figure 2. *Title page (at top) of* On the Origin of Species *(John Murray, London, 1859), and its author, Charles Darwin, by Thomas Herbert Maguire in 1849 (at bottom left), ten years before publication of the* Origin, *and in 1871 (at bottom right) in* Harper's Weekly. *(Bottom images courtesy of Cornell University Library.)*

3. A SHORT HISTORY OF EVOLUTIONARY BIOLOGY

Numerous individual scholars over the centuries – from ancient Greek philosophers to Leonardo da Vinci – have speculated about the age of the Earth or change in life through time. For most of human history, however, most educated opinions held that the Earth and its life were created relatively recently – perhaps just a few thousand years ago – essentially as we see it today. It is not surprising that this was so. It was the view supplied by most major religious traditions as well as common human experience.

This view began to change only in the early 19th century. Observed patterns in nature that had always been puzzling were becoming more and more troubling to many thinkers. Why, for example, are particular organisms found where they are, and not elsewhere? Why are different fossils found in different layers of rock? Why do fossils found in rocks of a region often resemble the living animals of that region more than any other? Why do embryos often look like adults of other animals? Why are what look like seashells preserved in rocks at the tops of mountains? Why do very different animals and plants nevertheless have numerous features in common? Scientists – or, as they were called at the time, "natural philosophers" – wondered whether living things could ever be explained by natural laws, such as Isaac Newton (1643-1727) had done so successfully for physical objects and phenomena.

Darwin's Two Arguments

Natural philosophers and other scholars of the late 18th and early 19th centuries who pondered these questions pursued two basic types of answers. Some tried to develop purely naturalistic (scientific) theories, but none were very successful. Most accepted some form of theological explanation, which said that organisms are the way they are because God made them that way. One of the most popular of these approaches was called the "argument from design," which argued that if an organism looked like it had been designed – for example, if its features appeared well suited for its environment and mode of life – then it *was* designed, by a supernatural designer. One of the most popular and influential defenders of this view was the English clergyman William Paley (1743-1805). In his 1802 book, *Natural Theology; or, Evidences of the Existence and Attributes of the Deity* (Figure 3), Paley argued that all features of organisms (as well as the rest of the world) were designed and sustained by God, insisting on "the necessity, in each particular case, of an intelligent designing mind for the contriving and determining of the forms which organized bodies bear." Paley started his book with what became a famous analogy:

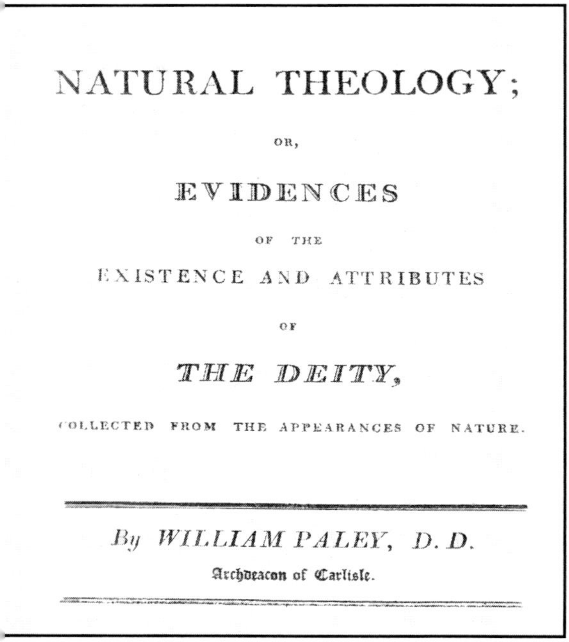

NATURAL THEOLOGY;

OR,

EVIDENCES

OF THE

EXISTENCE AND ATTRIBUTES

OF

THE DEITY,

COLLECTED FROM THE APPEARANCES OF NATURE.

By WILLIAM PALEY, D. D.

Archdeacon of Carlisle.

Figure 3. *Title page of the 1805 edition of William Paley's* Natural Theology. *Originally published in 1802, the book was Paley's last, but remained in print long after his death in 1805.*

> *In crossing a heath, suppose I pitched my foot against a stone, and were asked how the stone came to be there, I might possibly answer, that, for any thing I knew to the contrary, it had lain there for ever: nor would it perhaps be very easy to shew the absurdity of this answer. But suppose I had found a watch upon the ground, and it should be enquired how the watch happened to be in that place, I should hardly think of the answer which I had before given, that, for any thing I knew, the watch might have always been there. Yet why should not this answer serve for the watch as well as the stone? ... For this reason, and for no other ... that, when we come to inspect the watch, we perceive (what we could not discover in the stone) that its several parts are framed and put together for a purpose, e.g., that they are so formed and adjusted as to produce motion, and that motion so regulated as to point out the hour of the day ... This mechanism being observed ... the inference, we think, is inevitable, that the watch must have had a maker...*[2]

Paley's *Natural Theology* was among the favorite books of Charles Darwin (1809-1882; Figure 2) when he attended Cambridge University in the late 1820s. Darwin was still a conventional creationist like Paley when, soon after receiving his degree, he served for five years as naturalist on the round-the-world voyage of H.M.S. *Beagle*. During and immediately after the voyage, however, Darwin began to doubt Paley's explanation. He began to develop a nontheological explanation for what he had seen in places like South America and the Galapagos Islands, and spent more than 20 years gathering information that might support his ideas. In 1859, he published *On the Origin of Species by Means of Natural Selection, or the Preservation of Favoured Races in the Struggle for Life*, which by any measure ranks as one of the most influential books ever written.

Darwin later called his book "one long argument," but it was actually two interconnected arguments. In this book (which scientists commonly refer to in shorthand as "the *Origin*"), Darwin tried to do two things. First, he tried to convince his readers that evolution – he referred to it by the very useful term "descent with modification" – is the best explanation for the order, history, and diversity of life, by presenting a large compendium of evidence from all areas of biology. Second, he argued that a particular process, which he called **natural selection,** is the main cause of evolution (see Chapter 5).

Darwin succeeded in his first argument. Although he was by no means the first to argue for evolution. Darwin was the first to convince a large number of people that it was true, and it has been essentially universally accepted by the scientific community since approximately 1880. Darwin failed, however, in his second argument – natural selection was not immediately accepted by most scientists, and was widely accepted only in the 1940s. The differing success of Darwin's two arguments is not widely understood outside of biology, yet it is very important for understanding both the subsequent history of evolutionary science and the nonscientific controversies that continue to swirl around evolution to the present day. *The rapid acceptance of evolution (descent with modification) by scientists, but not natural selection, demonstrates that it is possible to accept that evolution has occurred without agreeing on the mechanisms by which it has occurred.*

Natural selection was Darwin's alternative to Paley's argument from design, and a radical departure from all previous speculations about possible causes of transmutation. It is an idea that runs counter to many of our fondest and most comforting hopes and beliefs, and for this reason has been difficult for many people to accept. More specifically, natural selection is a hard idea for many people to accept for at least four reasons:[3]

(1) Natural selection is mindless and purposeless. Features of organisms resulting from natural selection have the *appearance* of design but, Darwin said, they have been "designed" by a completely mindless process. Natural selection consists of nothing other than individual organisms possessing variations that enhance survival or reproduction replacing those less suitably endowed, which therefore survive or reproduce in lesser degree (see Chapter 5). There is no overarching direction provided by an intelligent designer. Nor are organisms assembled for any future purpose. They come to be as they are purely as a result of their living in the present. (Natural selection is thus a completely non-**teleological** theory. It is not directed by anything toward any particular goal.) The apparent harmony and beauty of the natural world are thus also without any grand plan or reason beyond survival and reproduction.

(2) Natural selection has no fixed direction. Natural selection is about survival and reproduction of organisms in their local environment, not about a greater direction toward cosmically "better," "higher," or more complex forms. Selection can in principle result in simplification and "degeneracy" as often as the reverse, depending on the demands of the environment. The overall direction of evolution, then, is not a result of any force driving or pulling life "upward" or anywhere else. It is a side-consequence of the interaction of directionless genetic variation and the patterns of environmental change. "Progress" in any general sense in the history of life is thus largely an illusion.

(3) Natural selection is completely materialistic. Previous thinkers had imagined that, if species did change, they did so due to frankly mystical or supernatural "vital forces" or simply because of divine will. Darwin, however, spoke only of random variation shuffled and sorted by natural selection (*i.e.*, the external environment). And he went even farther. Darwin made it clear that he believed not just that the human body was a result of this materialistic process, but also what he called "the citadel itself" – the human mind.

(4) Natural selection is creative. Many of Darwin's contemporaries reacted to the *Origin* by saying that they could understand how natural selection could *eliminate the unfit*, but not how it could *create the fit*. For this, they said, some other process, perhaps supernatural, would be needed. Yet the very essence of natural selection is that it is an engine of new forms, not just an executioner of old ones. As described further in Chapter 5, it does so by accumulating favorable variants, generation by generation, so that descendants come to have features that allow them to survive and reproduce at higher rates than their ancestors.

Evolution After Darwin

The **diversity** of life – the presence of a huge number of different kinds of living things – is one of the most conspicuous and puzzling aspects of life. Since the 1700s, scientists have used the word "species" to refer to these different kinds. A **species** is a group of organisms that "holds together," or looks the same, over some extent of time and space, and can be distinguished from other such groups. The details and causes of this coherence differ among different groups of organisms. In sexually-reproducing animals, most biologists today accept that a species is a group of organisms that reproduce among themselves but not with others. Despite the title of his most famous book, Darwin did not actually devote much attention to how and why new species come into being. Instead, he proposed a cause – natural selection – for the **transformation** of one species into another. Darwin did note that similar (and presumably closely related) species are often in different but adjacent places. He noted that the processes of making new species create fairly distinct products – each living in its own way. Yet he never put the pieces of the puzzle together into a coherent theory that would explain why there are so many kinds of living things.

Darwin and the other evolutionary scientists who worked in the half-century after 1859 were hindered by their lack of understanding of a field of biology that did not yet exist: **genetics** – the science of heredity.

Figure 4. *Gregor Mendel (1822-1884), the Austrian monk whose discovery of genes went unrecognized for almost 40 years.*

Ironically, it was during the 1860s, while Darwin was writing but unbeknownst to him, that an Austrian monk, Gregor Mendel (Figure 4), cracked the central problem of heredity. By carefully breeding peas in his garden, Mendel discovered that heredity is delivered to succeeding generations in discrete packages or units, later called **genes**. It is not diluted generation to generation, like different colors of paint being mixed together. Mendel's findings were largely unknown to the scientific community until 1900, when they were rediscovered. Then followed a confusing time, in which some scientists thought that, because genes are units of heredity that change only by noticeable change or **mutation**, evolution must occur by mutation alone – without Darwin's favored mechanism of natural selection. New mutations arise, suggested some of these theories, according to some unknown internal force within organisms, and this phenomenon directs the way new types of organisms evolve. Natural selection, if it happened at all, only resulted in the "weeding out" of the unfit. (None of these controversies had anything to do with whether descent with modification occurs, which by then no informed scientist doubted.)

At the same time, another development began to occur. Scientists working in laboratories in the new field of experimental genetics began to conclude that most characteristics of organisms are determined not by single genes (as Mendel had suggested), but by many genes, and that the **variation** produced by these numerous genes behaves mathematically in predictable ways. Several scientists began to realize that, if natural selection acted on small variations slowly over long periods of time, large-scale evolutionary change could be produced. Thus was born the field of **population genetics**. This research appeared to verify Darwin's conclusion that variation itself was not the direction-giving force in natural selection, rather it was the environment selecting from among the apparently "random" variants produced every generation that dictated the path of evolution (see Chapter 5, page 51).

At the same time that these laboratory studies were progressing, changes were also occurring in the field of biology concerned with naming and classifying the diversity of organisms: **systematics**. Scientists who name and classify kinds of organisms – systematists – were changing the way they viewed species. Instead of viewing each species as having a fixed set of characteristics, which if they varied slightly justified recognition of a new species, systematists began to see that species in nature are made up of **populations** – groups of interbreeding organisms that share many characteristics but also show variation in many characteristics. Species, these scientists realized, are made up of populations that expand and contract and show differing amounts of variation among their individuals at different times. It is this variation, said the systematists, which allows natural selection to occur, which is the cause of much of the change we see in evolution. Still there was no clear picture of why or how new species arise.

Ernst Mayr (1904-2004; Figure 5) was a German ornithologist studying the birds of New Guinea in the 1920s and 1930s. Papua New Guinea is just the largest of thousands of islands in the southwestern Pacific Ocean. On these islands, Mayr found many different forms of birds, which had been given different species names. Mayr realized that here was a natural laboratory for some of the theories that

Figure 5. *Ernst Mayr (1904-2005) was one of the major architects of the the refomulation of Darwinism in the 1940s known as the Modern Synthesis.*

Figure 6. *Speciation, the formation of two species from one. The original population of birds (at top) becomes divided into two parts. These two populations change over time (due to either natural selection or genetic drift). Eventually they are so different so that when they come back into contact, they cannot interbreed. They have become different species. (Bird illustration by Richard Kissel.)*

were being developed in the human-made laboratories of the geneticists. He hypothesized that birds had arrived on offshore islands from the mainland of New Guinea – due to storms or other accidents – and did not return. They bred and lived on these islands and formed new populations. Yet these new populations were isolated from those on the mainland – new birds did not arrive regularly enough to maintain genetic continuity. Natural selection, as well as other genetic processes, acted on a new population, which began to change in comparison with the original population. Eventually, the new population was so different genetically that if one of these birds did make it back to New Guinea, it would not interbreed with the birds there. A new species was formed – the process known as **speciation** (Figure 6). (Mayr was an expert on birds, which are sexually-reproducing animals. Other scientists working on other kinds of organisms, such as plants and bacteria, which differ in many ways from animals, found that new species arise in different ways, but agreed that essentially all of these processes of speciation could be understood in terms of population genetics.)

The final piece of the puzzle of a comprehensive evolutionary theory was provided by **paleontology**, the study of the fossil record. Prior to the 1940s, paleontologists did not generally favor natural selection as a mechanism of evolutionary change. It seemed too slow and insignificant to account for the grand and sweeping changes over millions of years that seemed so evident in the fossil record. Some form of internal forcing, some paleontologists said, must be responsible for the evolutionary sequences of magnificent horns and curved shells that appeared in the strata. In the 1940s, however, a few paleontologists had become acquainted with the theories of their colleagues in systematics and population genetics. In 1944, the American paleontologist George Gaylord Simpson (1902-1984) published *Tempo and Mode in Evolution*, a book in which he argued that, just as natural selection could be

Figure 7. *Paleontologist Stephen Jay Gould (1941-2002) argued for an expanded Neo-darwinism based largely on insights from paleontology. (Photograph courtesy of Kathy Chapman.)*

responsible for generating the diversity of life on the islands of New Guinea, so it could be responsible for generating the diversity of life visible in the fossil record. Fossils too could be seen as having been part of variable populations that waxed and waned under the influence of natural selection and undergoing speciation, essentially as Mayr had described.

This intellectual coming together of geneticists, systematists, and paleontologists to forge one coherent theory of how evolution works was called the **Modern (or Neodarwinian) Synthesis** because it synthesized findings from these three disciplines. It held that natural selection was the primary driving force behind evolution, selecting from among a large supply of small and largely randomly distributed variants every generation. It has been the dominant theory of evolutionary change since the mid-20th century and is essentially the view that is still presented in most textbooks.

Meanwhile, although Mendel had described how inheritance worked in general terms, the actual mechanism by which it worked remained unknown. In 1953, two biologists, American James Watson (b. 1928) and Francis Crick of the U.K. (1916-2004), described the detailed chemical structure of the molecule responsible for inheritance: deoxyribonucleic acid, or **DNA** (see Chapter 4). This discovery made it possible for scientists to begin to understand how evolution happens at its most basic level.

In the 1960s, some scientists began to challenge some aspects of Neodarwinism (again, without calling into question the concept of descent with modification). Geneticists, for example, noticed that a great deal of genetic variation appeared to be "neutral" with respect to natural selection – that is, it did not appear to affect the survival or reproduction of individuals. The issue of how such variation arose and was maintained, and how much evolutionary change it caused, became a major issue in evolutionary biology.

Similarly, in the 1970s and 1980s, several younger paleontologists, including Stephen Jay Gould (Figure 7), Niles Eldredge, and Steven Stanley, among others, began to argue that natural selection might not be so powerful after all, or that over geologic time other forces often diluted its influence. Based on their examination of fossils in layers of rocks, these paleontologists suggested that most species do not change very much once they arise, perhaps because their genes and body structure are tightly **constrained** within narrow bounds (see Chapter 5, page 51). They suggested instead that most evolutionary change occurs at the origin of new species – at events of speciation – when these constraints are broken down. In 1972, Eldredge and Gould called this idea **punctuated equilibrium**, which led to a reexamination of some of the ideas of the Modern Synthesis. Today evolutionary biologists and paleontologists accept some of the ideas of Gould and colleagues, while rejecting others.

Genes provide the "instructions" or "blueprints" for making an organism. The actual process of translating those instructions into the organism itself is called **development**, basically the growth of an organism. Just as builders modifying an old house are constrained by the existing house, land, and available materials (not to mention physical laws), so too organisms that might evolve are constrained to modify their existing genetic blueprint and to grow according to available materials and physical laws. This means that organisms are not necessarily constructed in a way that engineering would suggest is most efficient or effective. For example, if a species develops a new feature, like a shell or eye, it does so not by evolving a completely new tissue or structure, but by altering the structures it already has, frequently by co-opting a feature that serves an unrelated function (Figure 9). Biologists had known that this was an important part of

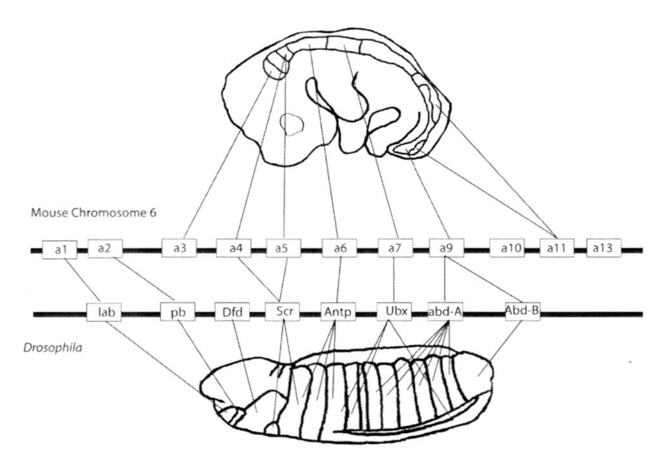

Figure 8. *An example of evo-devo. A mouse embryo (at top) shares a surprising number of genes with a fruitfly embryo (at bottom), which is further evidence that they share a distant common ancestor. Mice and fruitflies come to be different, however, in part because of the number of times that these genes are expressed as the animal grows, and the way that the genes interact with each other.*

the evolutionary story, but until fairly recently the details of development processes were largely invisible to them. In the 1980s, however, technical innovations started to make it possible for scientists to describe, or sequence, the chemical components of the genetic material, DNA, and this allowed in many cases for the determination of which genes provide instructions for the development of different structures or functions in the body. One of the major discoveries of this work was that very different organisms for example, fruit flies and people, share a surprisingly large number of genes. This suggests that it is not so much the presence or absence of discrete genes that controls the form of organisms, but how the genes that are present are expressed, that is, how and when they do their business within cells. These and other advances led to the establishment of an important new subfield of evolutionary biology known as evolutionary developmental biology, or **evo-devo**.

Evolutionary biology in the early 21st century is among the most active fields of science, and it continues to press forward feverishly on many frontiers, several of which are very briefly summarized in the chapters that follow. In just the past decade or two, research as

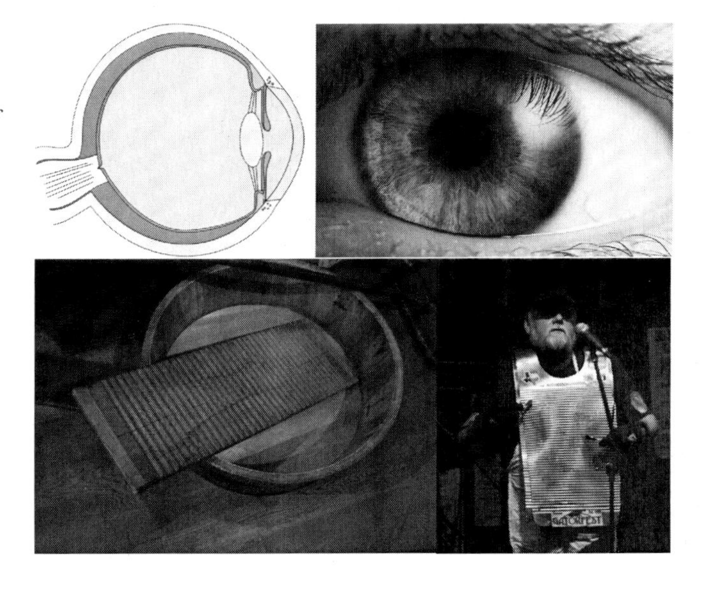

Figure 9. *The utility of intermediate stages in evolution. ID advocates suggest that Darwinism cannot explain the evolution of the human eye (at top) because it would not function unless all of its complex parts were present. Thus intermediate stages leading to it cannot have been favored by selection. But evolution often works by co-opting structures to different functions. The washboard (at bottom left) was designed for scrubbing clothes, but its "descendant" (at bottom right) evolved to be well suited for playing music. Many features of the washboard as a musical instrument originated for a completely different function.*

different as discovery of new fossils from around the "Cambrian explosion" of animal life more than 540 million years ago, the sequencing of the human genome, and the detailed description of molecular mechanisms of embryological development in numerous species have yielded rich and often unexpected results. As measured by any conventional indicator – number of scientists and students, popularity of courses in colleges and universities, productivity of research publications, popular interest among the general public and media, success of popular books, application to societal problems, and even research funding – evolutionary biology is exceedingly healthy as a scientific discipline. Two hundred years after Darwin's birth, and 150 years after the publication of its founding document, modern evolutionary biology contains almost exactly the right mix of well-established foundational conclusions, persistent controversies, new discoveries, and questions for it to remain among the most exciting fields long into the future.

Summary - History of Evolutionary Biology:

- In his 1859 book, *On the Origin of Species*, Darwin argued for two separate but related ideas – that evolution (descent with modification) is true, and that it is mostly caused by natural selection.
- Darwin did not invent evolution, but he assembled so much evidence that he convinced scientists that it is true.

- Darwin's proposed mechanism for evolution (natural selection) was not widely accepted by scientists until more than 60 years after his death.

- Darwin knew that inheritance happens, but he did not know why. The basis for heredity (the gene) was not widely known until the early 20th century, and its detailed chemistry (DNA) was not understood until the 1950s.

- The widespread acceptance of natural selection in the 1940s was a result of the combination of insights from several scientific disciplines, known as Modern Synthesis or Neodarwinism.

- Aspects of Neodarwinism have been challenged, modified, and expanded since the 1960s, especially by work in paleontology, molecular biology, and developmental biology, but most evolutionary scientists still think that natural selection is the most important, although not the only, cause of evolutionary change.

- Nothing discovered since Darwin's time has caused any serious scientist to doubt whether evolution (that is, descent with modification) is true, even while active scientific research and controversy continue about the mechanisms by which evolution occurs.

Figure 10. *Life's order (at top): a "drip tip" on a leaf in a Costa Rican rain forest, which functions to drain water off the leaf's surface. Life's diversity (at bottom left): a White-tailed Hillstar hummingbird* (Urochroa bougueri) *from the Andes of Ecuador. There are approximately 330 known species of hummingbirds. Life's history (at bottom right): the modern snail on the left* (Turritella exoleta) *from the Gulf of Mexico is similar to but recognizably different from the fossil snail on the right* (Turritella apicalis) *from a shell bed near Sarasota, Florida, dating from the Pliocene Epoch (approximately 3 million years old). (Photographs by W. D. Allmon, Robert Bleiweiss, and Kelly Cronin, respectively.)*

4. THE EVIDENCE FOR EVOLUTION

As discussed in the previous chapter, before Darwin there was no credible explanation for observations about living things, other than to say "God made them that way" or simply "they just are that way." Such explanations, however, are unsatisfying as science because they are not testable or falsifiable, and they do not lead to any greater understanding. Darwin was not the first to propose a scientific theory to explain the features, history, and distribution of living things. He was, however, the first to offer a purely materialistic, physical hypothesis that was closely argued, supported by abundant evidence, and appeared to explain a wide array of observations about organisms. Such a successful explanation provides powerful evidence that a theory is correct.

What are these "observations about organisms" that need explaining? How did Darwin's particular theory – evolution by natural selection – offer to explain them?

Life's order. Why do organisms have the arrangements, relationships, forms, geographic distributions, and patterns of similarity and difference that they do? Any acceptable scientific answer to such questions must explain one of the most obvious characteristics of living things: their frequently close "fit" or "suitability" to do what they do, the phenomenon usually called **adaptation**.[4] However, a successful theory must also explain features that are *not* especially suitable for the life of the organism that displays them. Evolution by natural selection (Darwinism) explains all of these observations by suggesting that organisms are related to each other in genealogical patterns ("family trees") and have evolved, mostly but not exclusively as a result of natural selection, to fit as well as possible – given the limitations of the genetic and anatomical structures they inherit – into their places in nature.

Life's history. The fossil record offers unavoidable evidence that life has had a long history of dramatic change. An important aspect of this history is an explanation for not just the success but also the frequent failure of organisms, because the vast majority of species that have ever lived on Earth are **extinct**. Evolution by natural selection explains life's history by suggesting that all life shares a distant common ancestor, which gave rise to all of the organisms that followed, and that the changes of life over the past 3.5 billion years are a result of the struggles of individual organisms to survive and reproduce in a spectrum of changing environments.

Life's diversity. The history of life is not just the history of change in form. Life on Earth – now and in the past – also shows astonishing variety. Evolution by natural selection argues that the history of theis incredible diversity of life – like the comings and goings of characters in an enormously long play – are the result of the struggles of individual organisms to survive and reproduce in a spectrum of changing environments.

Why are modern scientists so convinced that the explanations of life's order, history, and diversity offered by evolution by natural selection are so compelling? It is important to note that some observations about life are explained by (and therefore provide evidence for) Darwin's first argument – descent with modification – and others are explained by (and offer evidence for) his second argument – the mechanism of natural selection. We will divide our discussion into Darwin's two arguments, starting with descent with modification, and then (in the next chapter) consider its causes.

The evidence for descent with modification can usefully be grouped into six categories:[5]

(1) Observed small scale change. The first chapter of the *Origin of Species* is about pigeons. Darwin wanted to draw his readers' attention to an analogy between change under domestication and change in nature. For example, humans can see, or have seen in historic or recent prehistoric times, the change of wolves into Pekinese, grass into corn, wild cows into Herefords, wild ponies into quarter horses, jungle fowl into oven roasters, and many more. Since Darwin's time, we have watched fruit fly populations change in the laboratory from forms with wings into ones without, among many other changes. We can even create what would be called new species if they were found

Figure 11. *Observable small-scale change as evidence for evolution. Nineteenth century illustration (at left) showing some results achieved by selective breeding by English pigeon fanciers. Darwin used artificial selection by humans, which can change pigeons and other domesticated animals and plants, as a powerful analogy for the natural selection that causes evolutionary change in nature. Experiments on the common fruitfly (at right),* Drosophila melanogaster, *show that selection can create evolutionary change in the laboratory.*

in nature.[6] In domesticated plants and animals, we can actually watch changes occur generation to generation. We can also watch accidental human-induced changes, such as the spread of antibiotic resistance in bacteria, pesticide resistance in insect pests, or changes in the HIV virus that causes AIDS. In nature we can also see small-scale changes on short time scales. This kind of small-scale evolutionary change is often called **microevolution** (Figure 11).

Darwin's comparison with domesticated plants and animals was more than an analogy. Agriculture and other selective breeding activities are not *like* evolution – they *are* evolution, albeit evolution mediated by humans. Darwin argued that such observations on the short time scales available to us can be reasonably extended into much longer time scales to explain the history of all life. That is, he argued for applying extrapolation to questions about the history of life. As already discussed (Chapter 2, page 12), this kind of extrapolation is what scientists do all the time: observe or experiment on what they

Figure 12. *Biogeography as evidence for evolution: a hornbill (at left), from Southeast Asia, and a toucan (at right), from Latin America. These two birds share many superficial similarities, and live in similar ways in very similar tropical forests. Yet they are very different in most features and clearly belong to different groups of birds. Evolution explains this biogeographic pattern as resulting from two separate groups of birds being changed by natural selection in similar ways to become adapted to similar environments.*

have access to and then extrapolate or infer that the results are applicable to what they cannot directly observe or manipulate. If we reject this method for evolution, we must seriously question it in all other areas of science. Critics of evolution often argue that, although they accept microevolution (because they can't deny it because it happens almost before their eyes), they do not accept so-called **macroevolution**, which scientists believe occurs over thousands to millions of years, because such change cannot be witnessed within a human lifetime. In other words, such critics reject extrapolation as a valid scientific method as applied to living things. They do not understand (or do not admit) that they are rejecting the use of extrapolation in one field of science (in which they don't like the conclusions) but accepting it in others.

(2) Biogeography. Why are different kinds of organisms located where they are on Earth? Darwin wondered about why organisms should be distributed so particularly over the Earth's surface; what logical process could create such a distribution? There are two principal types of biogeographic distributions that provide compelling evidence for evolution:

(a) Similar habitats have different species. Coral reefs in the Red Sea and in Jamaica are both built largely of coral and algae. They both inhabit essentially similar conditions of temperature, depth, sunlight, and nutrient abundance. Yet of their thousands of species, almost none are in common. Why? The corals of Jamaica could probably survive in the physical conditions of the Red Sea, yet they are not there. Why not? Similarly, the North and South Poles are similar in climate and physical condition, yet penguins do not live at the North Pole and polar bears do not live at the South Pole. They could, but they don't. Why? Nearly all organisms are limited in their geographic range, so these are just a few of millions of examples (Figure 12). Why don't the same organisms live everywhere that they can possibly live? Evolution explains these patterns as the result of history. These organisms have evolved from ancestors that lived in different places. The corals of the Red Sea were changing over time independently of the corals in Jamaica, so that the species in these two places differ slightly in biological detail. Penguins evolved from ancestors that never made it to the northern hemisphere, and polar bears evolved from ancestors that never made it to the southern hemisphere.

(b) Nearby locations have similar species. Darwin was particularly impressed that the animals of the Galapagos Islands, located more than 600 miles off the west coast of South America, are most similar to those of the South American mainland. Why should this be so? Why are the organisms of the Caribbean islands most similar to those of the surrounding American mainland? Why aren't they more similar to, say, those of Africa, or Asia? This is a pattern typical of all organisms. Evolution explains such patterns as results of history – organisms evolving from an ancestor and moving to a nearby habitat.

(3) Comparative anatomy and evolutionary "vestiges." As mentioned above, when we look at living things – from bacteria to plants to animals – we cannot help but be impressed by how well "fit," "suited," or "adapted" they frequently are to their environments and ways of life. For example, an insect has color patterns that enhance its ability it to blend in with its background and thus escape predators, or a predatory mammal has teeth that allow it to quickly kill and consume its prey. These kinds of observations are the standard stuff of descriptive natural history, and (as discussed below) natural selection explains them very adequately. Yet organisms also show features that seem to be the reverse – to have no apparent function or adaptive value, or to have functions very different from those they were apparently originally built to perform. For example, ostriches and many other birds are flightless but nevertheless have wings; many vertebrates that normally lack hind limbs, such as snakes and whales, nevertheless have small bones inside their flesh under their tail; most humans can't wiggle their ears but all of us have ear muscles (Figure 13).

Such features, often referred to as **vestigial organs** or simply **vestiges**, are difficult to explain by means of theories like the "argument from design" (discussed in Chapter 3). Yet they are easily explained by (and are powerful evidence for) descent with modification. This explanation is not, however, for the reasons that are sometimes given. Some evolutionists have pointed to supposedly vestigial structures as evidence of evolution by arguing that an "intelligent" creator would not have designed living things with parts that seem fashioned for other purposes or that have little or no function at all. Although provocative, this is not an adequate explanation, because science has no way to test what would or would not be sensible from a supernatural creator's point of view. Furthermore, features that might at first seem to have no function could later be shown to have functional importance.

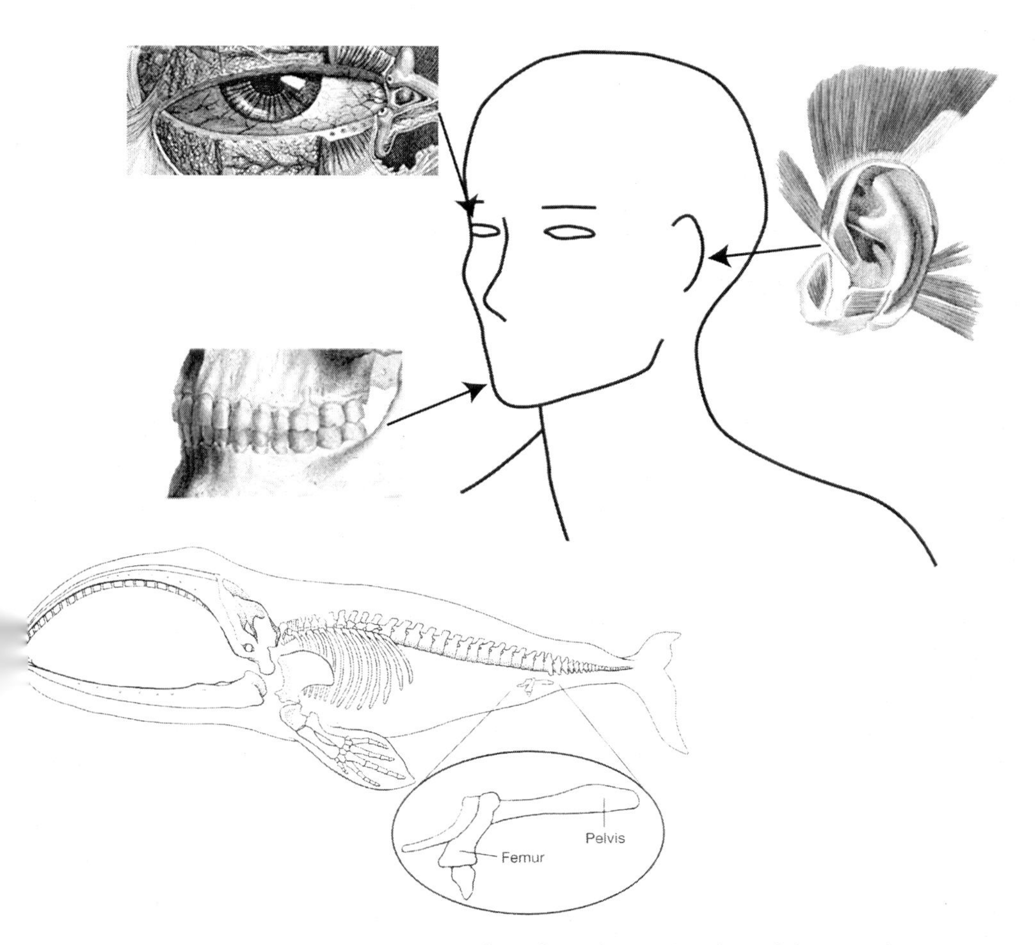

Figure 13. *Comparative anatomy as evidence for evolution. (1) Three of the many features of modern humans (at top) that do not appear to be adaptations to their present use, and which evolution interprets as vestiges left over from ancestors who used them for different purposes. (2) The skeleton of the modern Right Whale (at bottom) includes two pairs of small bones suspended below the tail. These bones appear to have no function today. Evolution explains them as vestiges of the hind limbs of whales' land-living ancestors of 50 million years ago.*

Vestiges are evidence for evolution because they often show a pattern across different species similar to what would be predicted if they were inherited from a common ancestor, rather than designed from scratch to meet the needs of just that one species. It isn't just that organisms have features that look like ad hoc or jury-rigged solutions. It is that they have features that look like they were jury-rigged in a particular sequence – through time as represented by fossils or in the classification of modern species. Organisms bearing such vestiges (and all do), therefore, do not appear to have been designed and built independently to meet the challenges of life with optimal engineering solutions, rather they appear to have been built to respond as well as they can to changing environmental demands by altering whatever was present at the moment. The process of comparing species to document these patterns of similarity is called **comparative anatomy**.

This question of the origin of vestigial structures is particularly interesting when we look at the earliest stages in the lives of organisms – the form of their embryos. Why do embryos look as they do? Why do they pass through the stages that they do? Why do human embryos have tails and gills? Why do bird embryos have teeth? Why do snails start out untwisted and then twist (and then sometimes untwist)? There are innumerable other examples.

Evolution explains such features – in embryos and adults – as the traces of history. According to evolution, organisms have many features that do not make sense in terms of their current function, but were inherited from an ancestor that used them in a different way. We all have ear muscles, for example, because a distant ancestor used them to move its ears.

(4) Fossils. The fossil record provides perhaps the clearest evidence for evolution, for it offers clear indications that organisms have changed through time and that there were different assemblages of species living at different times in the Earth's history. Critics of evolution frequently say that the fossil record contains no "transitional forms," presumably implying that the different fossils not linked by such forms were separately created This is simply incorrect. Literally thousands of fossils are known that are intermediate in their body form between other younger and older fossils.

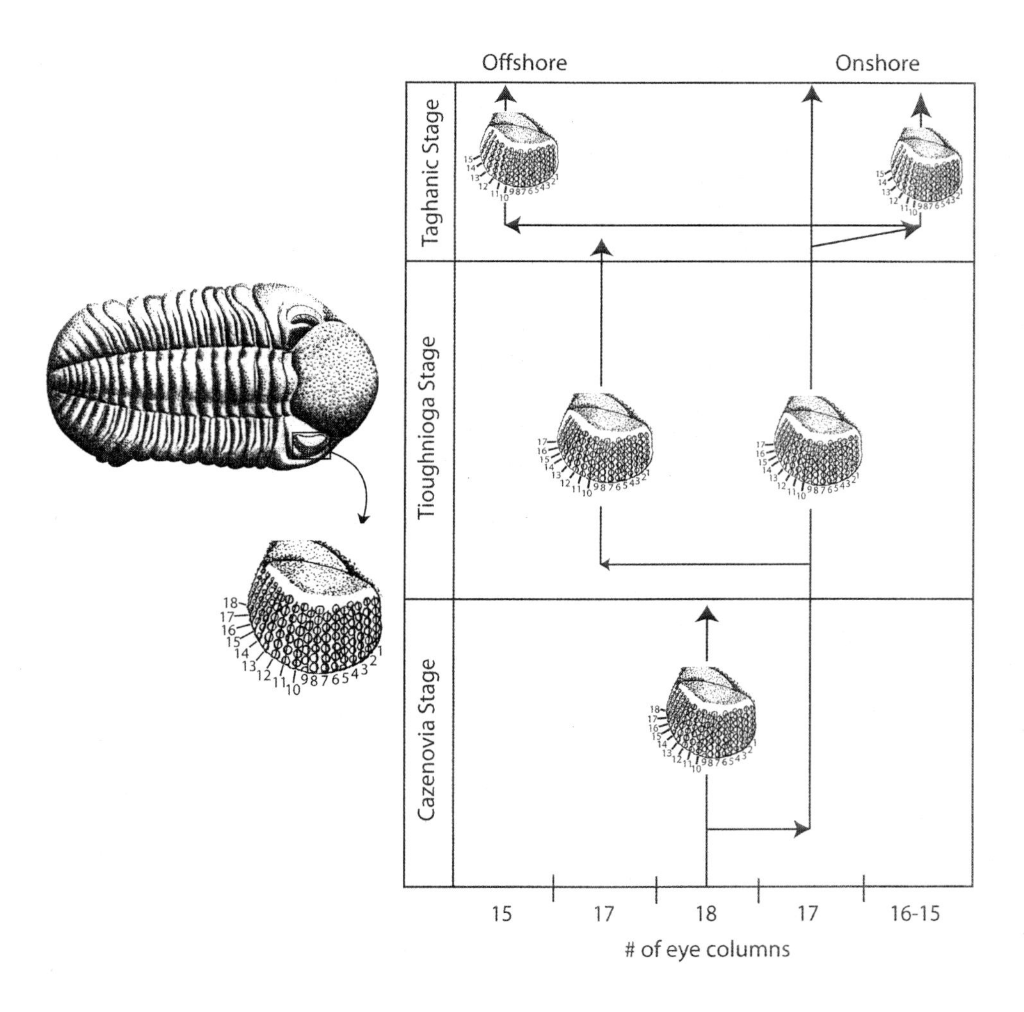

Figure 14. *An example of microevolution, or small-scale evolutionary change, in the fossil record. Fossils of the trilobite* Phacops rana *are found in rocks from the Devonian period (about 400-360 million years ago) in New York and Ohio. The lenses of the eye in these trilobites (above) are arranged in columns, which can be numbered. In a family tree of several forms of this trilobite (at right), fossils from lower (older) layers show different numbers of lenses in the compound eyes than fossils from upper (younger) layers. These trilobites inhabited a shallow sea that covered much of eastern North America in the Devonian Period. (Figure modified from N. Eldredge, 1985,* Times Frames: The Re-Thinking of Darwinian Evolution and the Theory of Punctuated Equilibria, *Simon & Schuster.)*

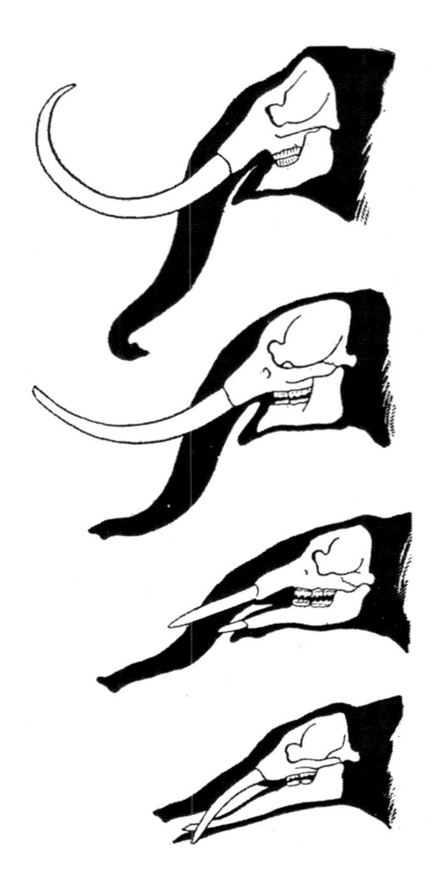

Figure 15. *An example of macroevolution, or large-scale evolutionary change, in the fossil record. Skulls of elephant-like animals from rocks of the Tertiary period (approximately 55 to 5 million years ago) in the midwestern region of North America look very different from modern elephants in the lowest, oldest rock layers, and increasingly similar to modern elephants in higher and younger layers. (Figure modified from W. B. Scott, 1913,* A History of Land Mammals in the Western Hemisphere, *MacMillan.)*

Fossil evidence for evolution can be divided somewhat arbitrarily into "large scale" and "small scale" patterns (Figures 14-15). In both cases the interpretation is the same: sequences of fossils appear in the rocks that are most easily interpreted (by reasonable extrapolation) as ancestor-descendant series, undergoing change in appearance. Recognizing that stacked layers of sedimentary rocks represent a span of time for deposition, it is difficult to understand the patterns in any other way. The fossils in these sequences are similar to frames in a cartoon or motion picture. The images in these frames do not actually move; we infer motion when we look at them in a series. More frames per second makes the motion look smoother, and fewer frames per second makes the motion look less smooth and more discontinuous or "jerky."

"Large scale" patterns are those that span tens or even hundreds of millions of years. Although there can be gaps of millions of years between rock layers that contain somewhat different-looking fossils, the fossils in successive layers look undeniably more similar than do fossils in more widely separated layers. It is possible that the organism in each layer was created separately and destroyed, to be succeeded by a newly created form that was just a little different. Evolution explains these patterns, however, as results of changes occurring during ancestor-descendant transitions.

"Small scale" patterns are sequences of fossils very closely spaced in layers of rocks, showing small but consistent patterns of difference among one another. These patterns would be even more difficult to explain in any way other than evolution, such as with a series of separate creations. They are similar to a cartoon with many frames per second – the inferred motion is highly believable because the changes between each one are so small. This is not something occasionally seen in just a few cases. They have been documented thousands of times by millions of fossils, collected over more than 200 years for scientific study and commercial applications such as petroleum geology. The consistency of these patterns is what gives them great value and provides us with confidence in our interpretation.

(5) Classification. Life is not arranged randomly and living things do not form a "smear" of variation. Organisms show patterns of similarities and differences that allow us to arrange them into groups, and to arrange these groups into groups, and so on. This hierarchical pattern is easily explicable by postulating that groups of organisms are connected by genealogy (Figure 16).

For example, all cats (lions, leopards, domestic cats, etc.) can be grouped together on the basis of their many shared features (such as retractable claws). Dogs and their kin (wolves, etc.) can be similarly grouped together on the basis of shared features (such as aspects of their teeth). However, cats and dogs can also be grouped together with other groups such as bears, seals, and weasels, because these groups share many features (for example, they are all carnivores and have numerous characters in common related to this, such as sharp claws and teeth). Carnivores are mammals, that is, can be grouped with other

Figure 16. *Classification as evidence of evolution. When we try to classify modern carnivorous mammals (and all other living things) into groups, we end up with a group-within-group, or hierarchical, pattern. Evolution explains this as a result of a branching family tree, in which groups that shared a more recent common ancestor share more similarities with each other than with groups with which they have a more ancient common ancestor.*

mammals, such as elephants and rodents, and so on. This "group-within-group" arrangement is explained by evolution, suggesting that the different kinds of cats share a more recent common ancestor with each other than any of them do with dogs, but dogs and cats share a more recent common ancestor than either group does with mice, etc.

(6) Genetics. All organisms on Earth contain RNA (ribonucleic acid) and almost all contain DNA (Figure 17). Virtually all organisms use exactly the same coding mechanism to read instructions from their genetic material – the same pattern of chemical subunits of DNA code for the same amino acids in oak trees and whales and people and mushrooms. Furthermore, all of the amino acids in living things on Earth are arranged the same way – all are "left-handed," even though "right-handed" forms also exist and have essentially identical chemical function. Why is this? Evolution explains this pattern by suggesting that all living things on Earth share a single common ancestor that contained RNA and left-handed amino acids, and this ancestor passed these features on to all of its descendants.

It is notable that Darwin did not know anything about genes or DNA when he wrote the *Origin of Species* in 1859. The existence of genes was not widely known until the early 20th century and the structure of DNA was not discovered until 1953. Yet the structure and function of genes and DNA are fully consistent with the hypothesis that organisms share a common ancestor and have changed through time – that they have evolved. In this sense, genetics has been an independent test, and confirmation, of the idea of evolution.

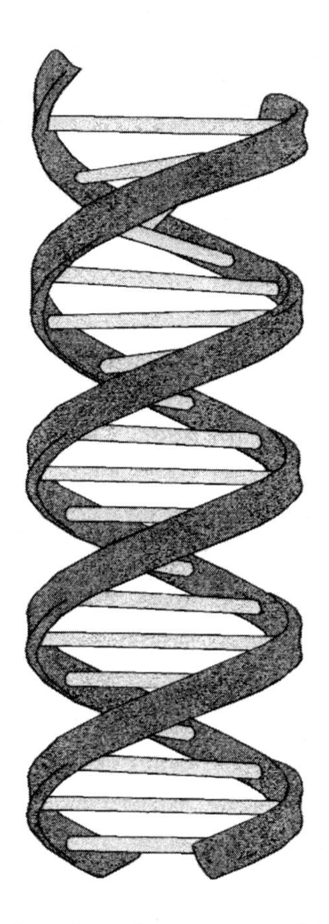

Figure 17. *Genetics as evidence for evolution. Darwin did not know how inheritance worked when he wrote* On the Origin of Species *in 1859, but his theory of evolution predicted that organisms would show patterns of genetic similarity that matched other evidence of their common evolutionary ancestry. The structure of the genetic material, DNA, was discovered in 1953 and Darwin's predictions have been fully confirmed since then. Living things show numerous similarities in their DNA that match similarities in other features.*

Summary - The Evidence for Evolution:

- Any adequate scientific theory of biology must explain at least the observed order (including both adaptation and nonadaptation), history, and the diversity of life.

- Pre-Darwinian explanations for these observations were largely theological, notably including the "argument from design," with which Darwin was very familiar.

- Evolution (in the sense of descent with modification) is an explanation for many of these observations. Natural selection as a cause for evolution is an explanation for others.

- The evidence for evolution can be grouped into six categories: directly observable small-scale change, biogeographic distribution, comparative anatomy, the fossil record, classification, and genetics. Each of these categories includes countless compelling pieces of evidence that descent with modification is true.

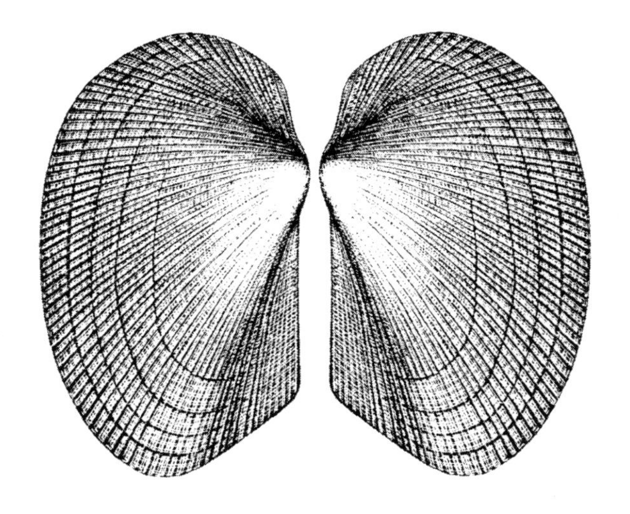

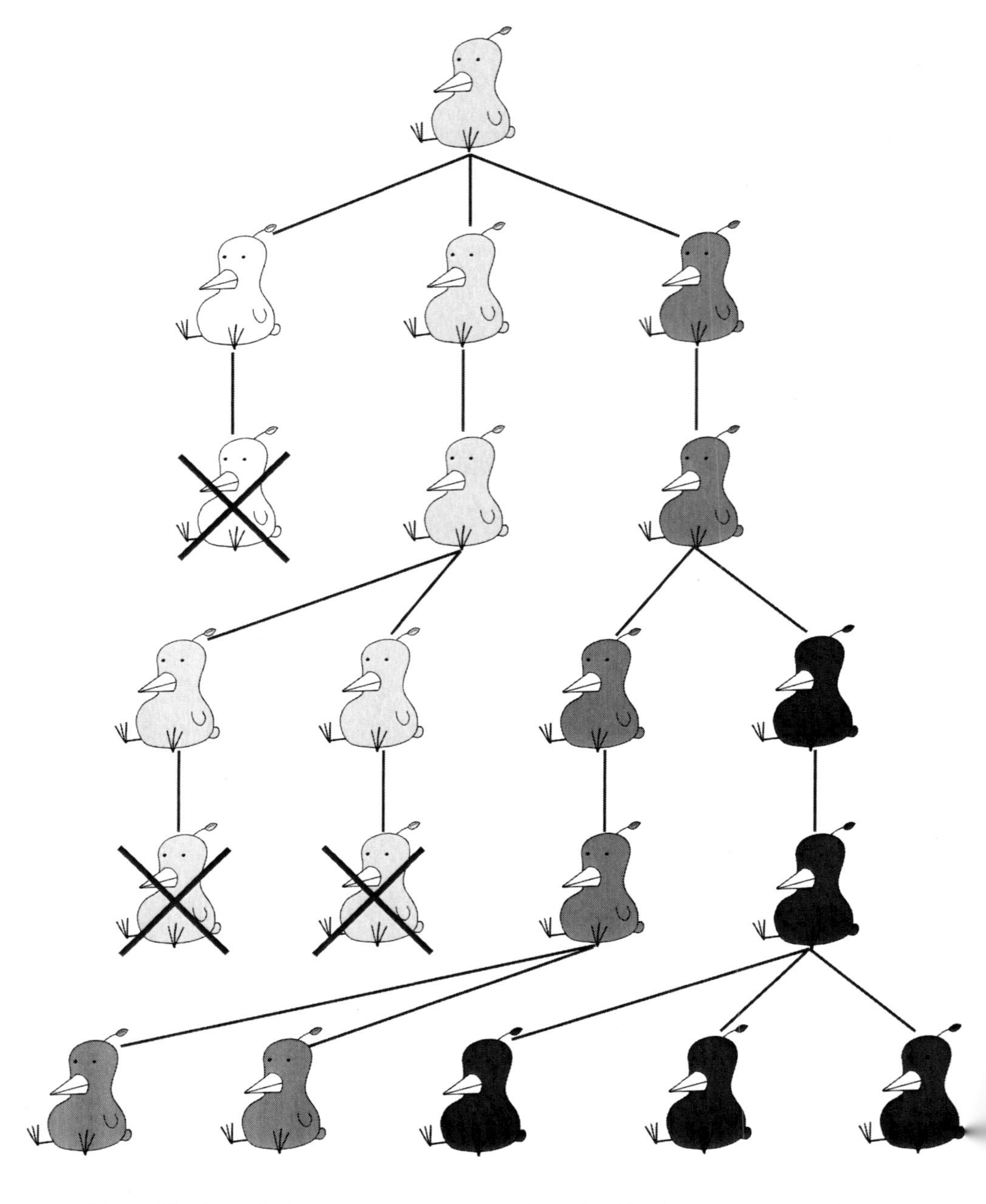

Figure 18. *Natural selection. In this cartoon, darker colored birds are more successful at surviving and reproducing than lighter colored birds, and so they increase their proportion in the population in later generations. Charles Darwin proposed natural selection as the primary cause for descent with modification in the* Origin of Species *in 1859. (Bird illustration by Richard Kissel.)*

5. CAUSES OF EVOLUTION

Evolution clearly has a variety of causes or mechanisms. Since the 1940s, scientists have generally agreed that the single most important cause of evolution is natural selection. Many evolutionary biologists, however, also think that other mechanisms in addition to natural selection are important in causing evolutionary change.

Natural Selection

Natural selection is a process proposed by Darwin in the *Origin of Species* to account for evolutionary change. He believed that it was the single most important, but not the only, mechanism for evolution. Using the analogy with artificial selection in domesticated animals and plants, Darwin described his new theory this way:

> *Let it be borne in mind how infinitely complex and close-fitting are the mutual relations of all organic beings to each other and to their physical conditions of life. Can it, then, be thought improbable, seeing that variations useful to man have undoubtedly occurred, that other variations useful in some way to each being in the great and complex battle of life, should sometimes occur in the course of thousands of generations? If such do occur, can we doubt (remembering that many more individuals are born than can possibly survive) that individuals having any advantage, however slight, over others, would have the best chance of surviving and of procreating their kind? On the other hand, we may feel sure that any variation in the least degree injurious would be rigidly destroyed.*

This preservation of favourable variations and the rejection of injurious variations, I call Natural Selection.[7]

Natural selection can be seen as a process involving four observable characteristics of all living things, and two processes that will unavoidably result from these characteristics. The four characteristics are:

(1) **Variation.** All organisms vary. There are no two individual organisms on Earth that are identical, not even identical twins.
(2) **Inheritance.** At least some of this variation is inherited.
(3) **Overproduction.** In all species, more offspring are produced than will survive to adulthood.
(4) **Fitness.** In all species, there are consistent relationships between particular inherited traits in individuals and the ability of those individuals to survive or reproduce.

These four characteristics produce two inevitable results:

(1) **Struggle for existence.** Because not all offspring are guaranteed the right to grow to adulthood, there is what Darwin called a "struggle for existence,"[8] in which every individual must strive to survive and reproduce. That is, every individual must compete with others for the resources it needs to survive and reproduce, and many individuals will not be successful in this struggle.
(2) **Change in genetic composition.** On average, those individuals with heritable traits that confer some advantage in the struggle to survive and reproduce will leave more offspring. This will cause a high percentage of individuals in later generations to possess these advantageous traits.

Natural selection starts with heritable variation, which is its "raw material." Darwin did not know the source of this variation (that would have to wait for the rediscovery of Mendel and the work of Watson and Crick), or the reasons for its inheritance, but he observed that it was usually abundant and in many directions. That is, the variation observed within a species was usually much broader than the direction of evolutionary change in that species. It is therefore said to be "random with respect to the direction of change."

Use of the word and concept of randomness has produced a lot of misunderstanding. Critics of natural selection have claimed that it requires that life is governed completely by "chance." One contemporary of Darwin said that it was the "law of higgledy-piggledy." But this is not what Darwin said at all. Natural selection is, in fact, anything but random. It is a highly determined process. The variation (ultimately produced by mutations) is in multiple directions, but the environment selects only some of these variations, accumulates them generation after generation, and therefore shapes the eventual direction of change in the population. Thus, the direction of evolution (*e.g.*, whether horses get bigger, birds get bluer, or shells get thicker) by natural selection is provided by the environment, not the underlying genetic variation.

An individual organism's **fitness** is the rate of increase of its descendants in later generations. Differences in fitness among individuals are average differences in the probability of reproductive success that are due not to chance, but to some characteristic difference between them – to one or more of those advantageous traits. Natural selection can therefore be defined as *any consistent difference in fitness (i.e., survival and reproduction) among groups of organisms which leads to changes in their genetic characteristics.*

It is important to understand what natural selection does *not* do. It does not guarantee perfection of adaptation, or even adaptation at all – it can only select from among the variation that is provided to it. The supply of variation can therefore act as a **constraint** within which natural selection can work. Natural selection also does not produce absolute improvement or "progress," only relative improvement within a local environment. As the environment changes, so too does the success of different individuals and lineages. The winners in one environment might become the losers as the environment changes and different traits become favorable. Natural selection does not necessarily lead to greater complexity – it can also produce simplification if this is advantageous in a particular environment.

It is common for evolutionary biologists to say or write that natural selection "acts" on organisms or genes, or that it is a "force" in evolution. Such language, however, is only convenient shorthand. Darwin himself wrote that "The term 'natural selection' is in some respects a bad one as it seems to imply conscious choice ... For brevity sake I sometimes speak of natural selection as an intelligent power: – in the same way as astronomers speak of the attraction of gravity as ruling the movements of the planets."[9] Darwin correctly recognized that this kind of scientific slang can be misleading – if it is not carefully qualified, listeners or readers can easily get the impression that natural selection is some kind of discrete entity or power. It isn't. Natural selection does not "act" and is not a "force" – it is even somewhat misleading to refer to it as a "process." Natural selection is a *result* of heritable biological differences among individuals interacting with the local environment, which can lead to genetic change in populations and species.

Other Causes

As Darwin himself recognized, natural selection is not the only mechanism or cause of evolutionary change, but he argued that it was likely to be the most important. This conclusion was solidified in the 1940s when improved understanding of genetic inheritance, the population structure of wild species, and the nature of the fossil record was combined into what became known as the Neodarwinian Synthesis. At the core of this view were the ideas that (1) most variation has either a positive or negative effect on survival or reproduction; (2) variation is abundant and in all directions – or at least not preferentially in any particular direction – and (3) natural selection gradually changes populations over long periods of geological time. Causes other than natural selection that are today taken seriously by a significant number of evolutionary biologists include several that do not follow these ideas.

Even at the height of the Synthesis in the 1950s, evolutionary biologists acknowledged that selection was not responsible for everything. Some variations, for example, seem to have no effect on survival or reproduction, and so are "invisible" to selection. Characteristics that affect reproductive success control the direction of change in the population, whereas these "neutral" or "nonselected" variants increase or decrease essentially randomly. And in small populations, however, some of these variations could decrease until they disappear altogether, eliminating these variants from a population wholly by chance. This is called **genetic drift** and was held to be a complement, rather than a replacement or competitor, to natural selection. Like natural selection, genetic drift is a mechanism of evolution that leads to changes in the genetic makeup of a population, but unlike natural selection, it is truly a random process driven wholly by chance events.

Beginning in the 1970s, some paleontologists began to point to two patterns in the fossil record that they claimed had been neglected by Neodarwinism, and they claimed that these patterns (1) called for a reassessment of some elements of the Synthetic theory and, perhaps (2) suggested the need for additional alternative hypotheses of evolutionary cause. The patterns were those that the influential paleontologist George Gaylord Simpson had selected for the title of his contribution to the Modern Synthesis: **tempo** and **mode**. Tempo referred to the rate of evolutionary change. Mode referred to the process – either the wholesale transformation of an entire species, or its breaking into two – which is called speciation.

The younger generation of paleontologists took aim at new understandings of both tempo and mode. When did evolutionary change occur, at what rates, and how did this correspond to transformation versus speciation? The apparent lack of gradual change in many fossil sequences – which came to be called **stasis** – challenged the Neodarwinian emphasis on gradual change within evolutionary lineages driven mostly by natural selection. It also suggested that the variation that Neodarwinism had assumed was effectively never itself a directional factor might in fact be more structured than both Darwin and the Neodarwinians had thought. Second, the apparent coincidence of evolutionary change in physical form with the occurrence of the splitting of lineages (speciation), after which little change occurred (the pattern known as punctuated equilibrium; see Chapter 3), also seemed to throw the power of natural selection into question. Could it be that large-scale evolutionary **trends** over millions of years could be caused by natural selection, but only during brief intervals associ-

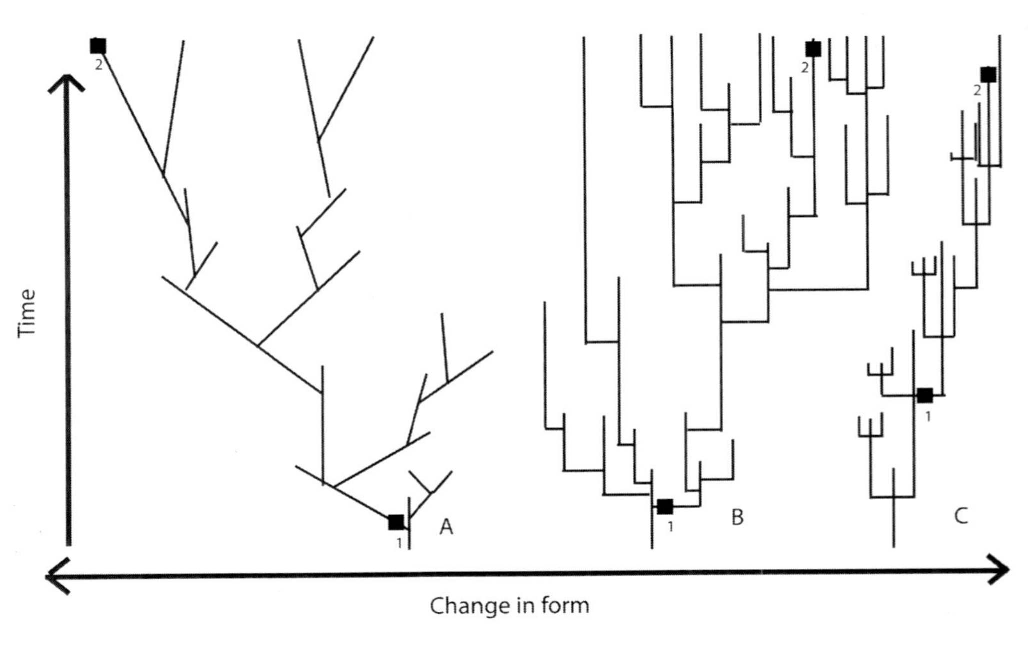

Figure 19. *Evolutionary family trees as indicators of evolutionary tempo and mode. Each line on the trees represents a species. Tree A shows gradual change in form within ancestor-descendant lineages. Evolutionary trends are due largely to this mode of change, which might reasonably be attributed to natural selection. Trees B and C, however, show change occurring between rather than within lineages, that is, at times of lineage branching or speciation. Evolutionary trends in these cases cannot be due to natural selection acting gradually within lineages, but are the result of sorting among species. Tree B suggests that a trend might be due to more speciation in one branch of the tree; tree C suggests that it might be due to more extinction in a branch than in another.*

ated with speciation? Or, some of the paleontologists suggested, could such very long-term macroevolutionary patterns be caused by differences in the rates of extinction and origination of species, rather than mostly by the birth and death of individuals within species? This has been called the **species sorting** theory of macroevolution. Exactly how important all of these various processes are, and how they interact with selection in the dramatically changing environments of the last half-billion years and more, are currently extremely active areas of research in paleontology (Figure 19).

At around the same time that these younger paleontologists were re-examining the implications of tempo and mode, molecular and developmental biologists were combining forces to discover the secrets of how tiny, simple, single-celled embryos turn into large, complex, multicellular adults in such great variety. As already mentioned in Chapter 3, these students of what became known as evo-devo soon reached conclusions similar to some of the paleontological hypotheses regarding the nature of the variation occurring in each generation and on which natural selection depends. Evo-devo scientists discovered that *not* all variations were possible, by a long shot. Certain similar variations were evidently "built-in" to the genetic material of species as different as flies and humans, and these variants provided a limited array of possible material on which natural selection could act. Evo-devo quickly became one of the largest and fastest moving areas of biological and evolutionary science. New and surprising results appear regularly, and it is clearly one of the places from which we will continue to learn a great deal about the causes of evolution.

Summary - Causes of Evolution:

- Darwin believed natural selection to be the main, but not the only, mechanism of evolutionary change. This is also the predominant view among evolutionary biologists today.

- Natural selection can be seen as a process involving four observable characteristics of all living things (variation, inheritance, overproduction, and consistent relationships between characteristics and reproductive success [fitness]), and two processes (struggle for existence and genetic change) that will unavoidably result from these characteristics.

- Major causes of evolution other than natural selection include genetic drift, species sorting, and developmental constraint. The second and third of these are areas of extremely active current research.

6. GEOLOGIC DATING, OR "HOW DO YOU KNOW HOW OLD IT IS?"

Although geological dating is not strictly part of evolutionary biology, knowing how old rocks are is an important part of using the fossils that they contain to study the history of life, including evolution. Critics of evolution often object to one or another aspect of geological dating. Thus an understanding of how geologists know the age of rocks and fossils is very useful for understanding evolution.

Answering the question "how do you know how old it is?" in geology requires two separate steps. They correspond to two different senses of how we tell how old anything is in our everyday experience. When asked how old an object or person is, we can answer either with a number or by comparing it to something (or someone) else. Thus, you might say "I am older than my brother" or "my friend's car is older than mine." This is called **relative dating**, because the age of something is stated relative to the age of something else. We can also give an age in numerical units, such as days, months, years, etc. This is usually called **numerical** (or, misleadingly, absolute) **dating**.

Relative Dating

Relative dating in geology makes use of two very reasonable (and testable) assumptions, and one common observation.

Assumption 1: Superposition. Rocks that formed from sediment (mud, sand, gravel) are called sedimentary rocks. Such rocks are usually seen to be arranged in stacks of layers or strata. These stacks are commonly called stratigraphic sequences. When we look at a stack of sedimentary layers, we can ask which layers are older, that is, which formed first? By reference to our common experience with such things as stacks of magazines on the living room floor, we can suppose that, in the absence of evidence to the contrary, the oldest layer in a stack of rocks is at the bottom, and that the youngest is at the top. This principle of geological reasoning is called **superposition**.

Observation: Succession of fossils. When we examine sedimentary rocks, we often find that they contain **fossils**. Fossils are the remains or traces of organisms from the geological past that are preserved in rocks. (We think that fossils were once associated with living organisms because they resemble organisms alive today.) When we look at fossils in stacks of sedimentary rocks from many places, we notice that different kinds of fossils occur in different layers and that the order of the various kinds of fossils from bottom to top is always the same, even in different places. This is called **biological succession**.[10]

Assumption 2: Correlation. When we look at fossils in stacks of rocks in different places, we make the reasonable assumption that, in the absence of evidence to the contrary, layers containing the same fossils in separate locations are similar in age. This is called **correlation**. The reason we can generally make this assumption with confidence is the extreme consistency of geological succession of organisms and other geologic features, and a large number of independent geologic observations (layers of volcanic ash, storm deposits, and many others) that show the same pattern.

These assumptions and this observation allow us to construct series of fossils that occur in different layers of rocks (Figure 20). As we

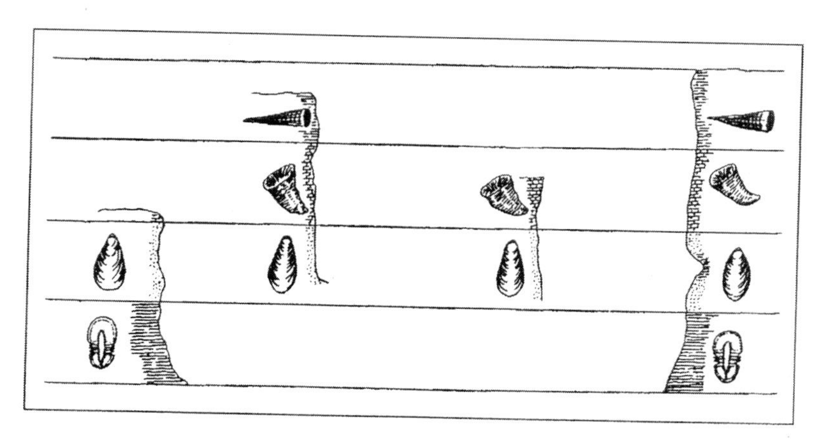

Figure 20. *Using fossils for correlation and relative dating of rock layers. Different fossils occur in different rock layers. If we assume that layers on the bottom of a stack of strata are older than layers on the top (superposition), and that similar fossils are the same age in different places, we can use fossils to correlate from place to place and establish a series of relative ages. This also creates new hypotheses that can be tested as new localities are explored.*

travel to more and more places, correlating stratigraphic sequences of geological succession as we go, we construct a grand series of fossils, oldest at the bottom and youngest at the top (according to superposition). For convenience we divide the long series of fossils into sections and name them. The names are usually based on places at which rocks of that particular age were first well-studied and represent the interval of time during which a particular set of organisms existed. This series of names is the **Geological Time Scale**, the internationally accepted system for telling time in geology (Figure 21).

The names on the Time Scale are labels for different groups of fossils. The Devonian Period, for example, has fossils first found in Devonshire, England. These fossils and their succession are like the ones found, for example, in central New York State, and so geologists conclude that the rocks of this area belong to the Devonian Period. This conclusion has since been affirmed by other sorts of independent geologic information.

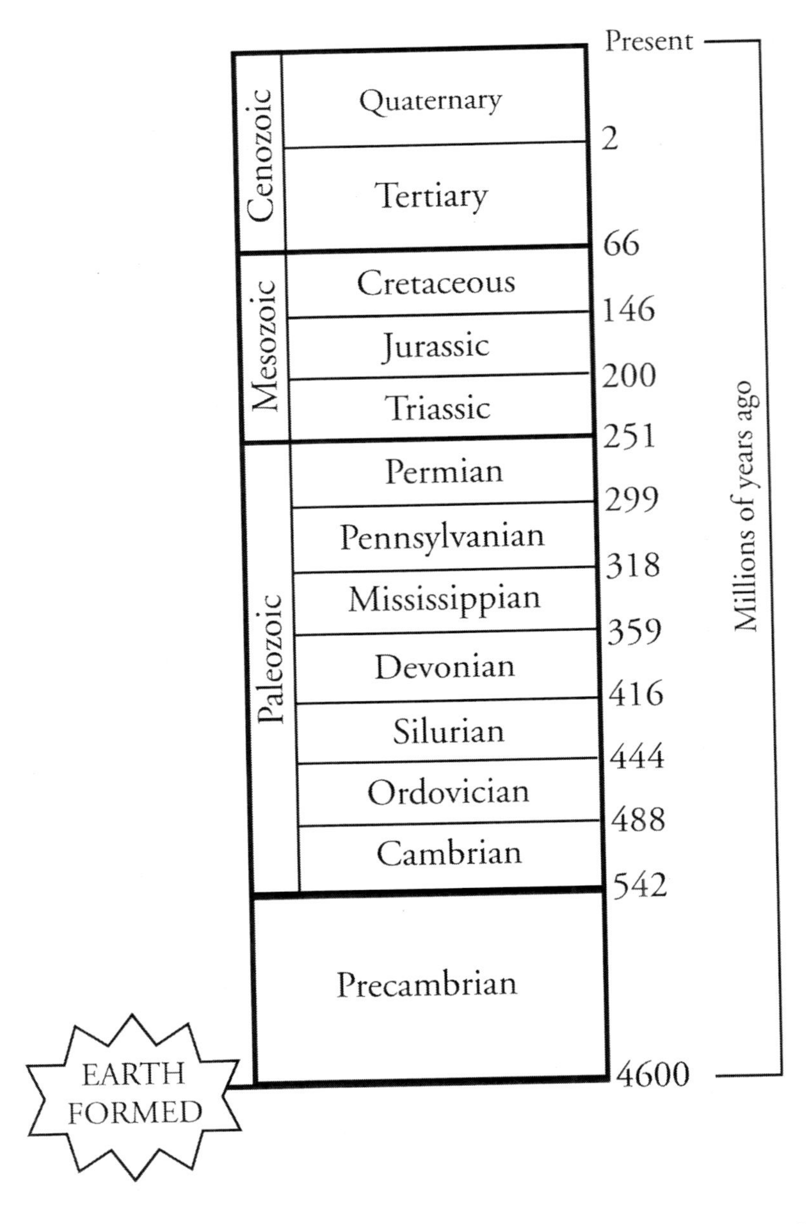

Figure 21. *The Geological Time Scale. The words in the time scale are really codes for groups of fossils that are found in different layers of rocks. The numbers are based on using radiometric dating on rocks that occur above or below particular fossils. These dates could change as refinements are made or new rocks are discovered. (From the 2004 version of the internationally accepted Time Scale.)*

Numerical Dating

The numerical ages of rocks in the Time Scale are not determined by fossils but by **radiometric dating**, which makes use of a process called **radioactive decay** – the same process that goes on inside a nuclear reactor to produce heat to make electricity. Radiometric dating works because radioactive elements decay at a known rate (measured in the laboratory). They act like ticking clocks, and let geologists measure how much time has passed since those elements were sealed into a particular mineral in a rock. Radiometric dating provides the numbers of years that are found on most versions of the Geological Time Scale. These numbers are revised occasionally, as more precise radiometric methods are developed.

Fossils themselves usually cannot be dated radiometrically. Radiometrically datable minerals usually only occur in volcanic rocks. Because fossils usually occur in sedimentary rocks, we must usually combine information from fossils and radiometric dates from rock layers above or below to answer the question "How do you know how old it is?" (Figure 22).

Although the numerical age of the Earth and the rocks and fossils we examine from its crust are not strictly a part of evolution, they are frequently included in discussions of the subject. This is because if the Earth is very young (say, 10,000 years old), as was widely believed centuries ago, there would simply have been insufficient time for the changes that have obviously occurred to have happened by naturalistic processes, and some kind of supernatural process would have to have been involved. Once it became clear (in the early 20th century) that the Earth was hundreds of millions of years old, evolutionists effectively stopped thinking about the topic, as this was more than enough time for the observed changes to have occurred by natural processes. The conclusion – reached by the mid-20th century – that the Earth is more than four billion years old did not affect evolution as much as it did our overall sense of place in the universe and its history. Indeed, the discovery that the Earth is extremely old – sometimes referred to as "deep time" – could be geology's most important contribution to human understanding.

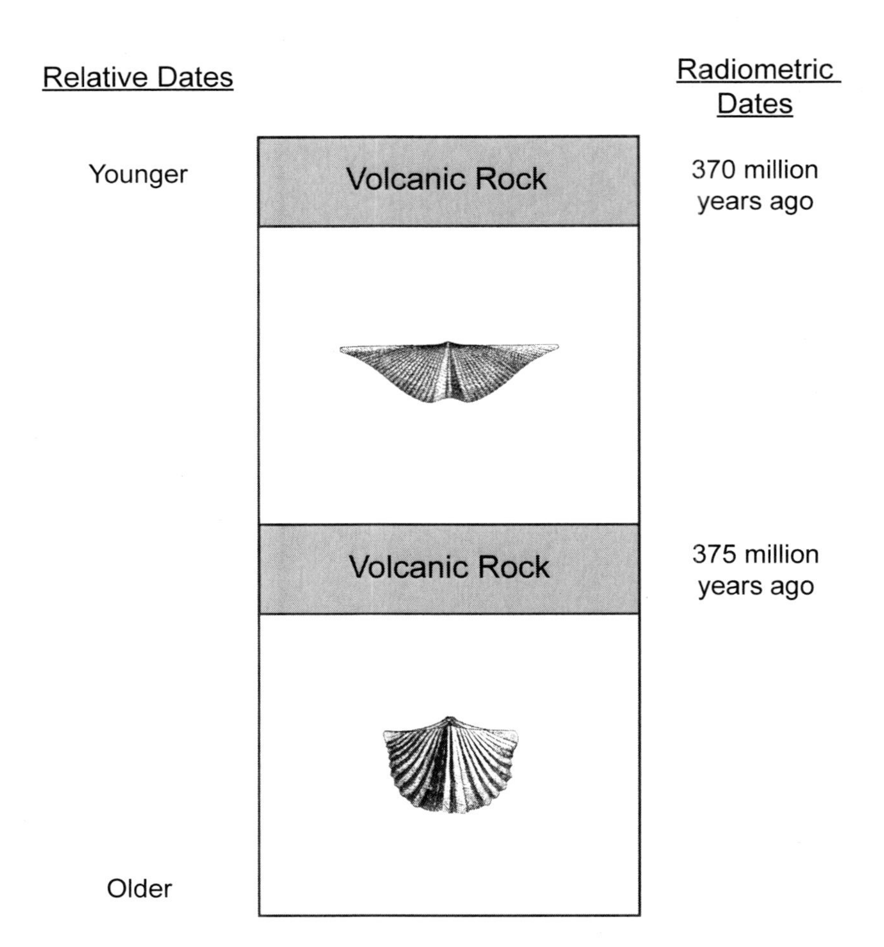

Figure 22. *Using radiometric dating to date fossils usually requires finding volcanic rocks interbedded with fossil-bearing rocks. Volcanic rocks contain mineral grains that include radioactive elements whose decay can be measured. The resulting date is older than fossils above the volcanic rock and younger than fossils below the volcanic rock. (It is important to note that such juxtaposition of fossils and radiometrically-datable layers provides an independent test for the validity of the assumption of superposition, which says that the oldest layers and fossils are on the bottom. Application of such tests almost always confirms that superposition is correct.)*

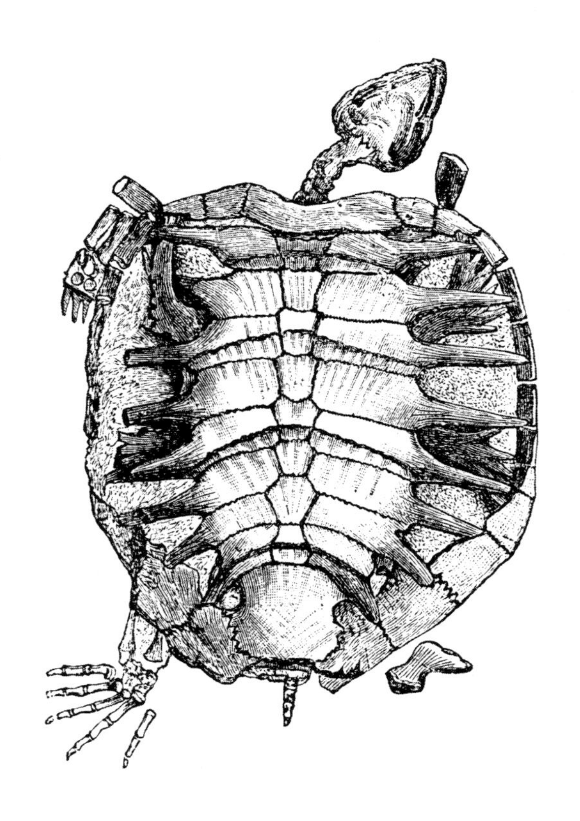

Summary - Geologic Dating:

- Determination of the age of rocks and geological phenomena can involve one of two different approaches: determination of relative age (expressed as older vs. younger) and determination of numerical age (expressed in years before present).

- Relative dating is accomplished by the application of two techniques: physical stratigraphy by the principle or assumption of superposition (which says that in a series of undisturbed sedimentary rocks, the oldest layer is on the bottom), and biostratigraphy by the principle or assumption of correlation (which uses the observation that different fossils occur in different layers and assumes that similar fossils are of similar age).

- The result of relative dating is construction of a composite sequence of rocks and fossils, which is a hypothetical but testable series of older-to-younger fossils and the rocks that contain them. This composite is known as The Geological Time Scale.

- The Geological Time Scale was assembled in essentially its present form by around 1830, more than three decades before the wide acceptance of evolution. Evolution is therefore obviously not required for the development or use of the Time Scale.

- Numerical dating is accomplished most often using radiometric dating, which uses the observed clock-like rates of radioactive decay of naturally-occurring elements to measure the amount of time that has passed since the minerals containing them formed.

- Radiometric dating was first done in the early twentieth century, following the discovery of radioactivity in 1896. This means that numerical dating is not required for establishment or use of The Geological Time Scale, which is based on fossils.

- Radiometric dating is most often performed on igneous rocks, which do not normally contain fossils. Assigning numerical dates to The Geological Time Scale, therefore, requires examination of occurrences of radiometrically datable layers (such as volcanic ash) above or below layers containing fossils.

- Scientists have high confidence in the conclusions of geological dating by relative and numerical techniques because they are tested and refined by normal geological work literally every day.

7. CREATIONISM

Creationism is the belief that the Earth and its life were created by a supernatural power. Prior to the middle of the 19th century, almost all individuals whom we would now call biologists and geologists were creationists. Darwin himself was a creationist when he sailed around the world on H.M.S. *Beagle* in the 1830s. What he observed on that voyage and learned afterward eventually convinced him, however, that supernatural creation was not an adequate scientific theory to explain the history, order, and diversity of life. Darwin published the *Origin of Species* in 1859. By the time he died in 1882, most biologists and geologists in Europe and the United States accepted that evolution was true (even though most did not accept Darwin's preferred cause of natural selection). Darwin is buried among royalty and heroes in Westminster Abbey – England's most honored place – because he convinced the scientific world that evolution had occurred.

A Brief History of Modern Creationism

Initial reaction to Darwin's book among the general public was mixed. Although it sold out its first printing on the first day, the overall reception was mostly cautious to negative across Europe and in the U.S., especially among religious leaders. As scientists began to accept evolution, however, loud public objections became fewer. Evolution was in many ways part of the general trends of "modernism" and "secularization" sweeping the Western world during the economically and technologically dynamic epoch of the late 19th and early 20th centuries.

In the wake of World War I, however, there was something of a backlash against modernity, especially in some areas of the United States. Laws were passed in several states making it illegal to teach that "man was descended from the lower animals." One such law in Tennessee was challenged by the American Civil Liberties Union in the famous Scopes trial, which took place in the summer of 1925 in the little town of Dayton, Tennessee. Although the popular impression of the Scopes trial (based in large part on the play and film *Inherit the Wind*), is that evolution triumphed, this is not true. Teacher John Scopes was convicted, fined $100, and the law stayed on the books until the late 1960s (Figure 23). Evolution was for the most part not taught in U.S. public schools for most of the middle of the 20th century.[11]

In the 1960s, motivated largely by the launch of the Sputnik satellite by the Soviet Union, American science education underwent dra-

Figure 23. *John Scopes (1900-1970) was a popular young school teacher in Dayton, Tennessee when he agreed to be the defendant in a case that would test the state law that prohibited the teaching of evolution. The famous "Scopes Trial" took place in Dayton in July 1925. After the trial, Scopes went on to graduate school at the University of Chicago and became a geologist. (Photograph from the Smithsonian Institution Archives.)*

matic improvements, and evolution was introduced into most science textbooks and classrooms. In 1968, the U.S. Supreme Court ruled in *Epperson v. Arkansas* that laws banning the teaching of evolution violated the First Amendment of the Constitution, which states that "Congress shall make no law respecting an establishment of religion," by prohibiting the teaching of a scientific theory for religious reasons.

In the late 1970s and early 1980s, encouraged by the nation's increasingly conservative political mood, critics of evolution began arguing for what they called "scientific creationism" as a scientifically valid theory that should be given "equal time" with evolution in public school science classrooms. These arguments for "scientific creationism" usually consisted of little more than miscellaneous objections to evolution or natural selection, many of which were presented via misrepresentation and selective or misquotation of evolutionists themselves. In 1981, in *McLean v. Arkansas*, Federal Judge William Overton ruled that "scientific creationism" was not science but a clear attempt to promote a particular religious view. This was confirmed in 1987 by the U.S. Supreme Court in *Edwards v. Aguillard.*

For the next decade or so, creationism was largely out of the national spotlight, but it was very active at the local level, for example, encouraging school boards to require teachers to read to their students misleading statements such as evolution is "only a theory" or to adopt watered-down text books. In the background, however, a new version of creationism known as **intelligent design** (ID) was being developed. ID is the idea that features of the physical universe and/or life can be best explained by reference to an "intelligent cause" rather than a natural process or material mechanism. Although (as discussed in Chapter 5) the concept of ID is not new at all, because it can trace its origins to the early 19th century or even earlier, beginning in the early 1990s modern ID developed several new approaches with the writings and activities of a group of politically conservative thinkers including Michael Behe, Phillip Johnson, William Dembski, Stephen Meyer, and John West.

These modern ID advocates argued that it was a scientific and non-religious alternative to Darwinian evolution, and many of its advocates resisted being grouped with other creationists. Advocates of ID argued that there are features of organisms that are "irreducibly complex," that is, they would not function if one element was removed, such as the human eye or the red blood cell. They argued that such features could not be produced by incremental additions via natural selection because intermediate stages would not be viable, and that such features could therefore only result from the actions of a supernatural designer. They favor "teaching the controversy" between ID and Darwinian evolution in science classrooms.

Despite its claims to scientific credibility, modern ID was fraught with problems. For example, although advocates of ID stated that it was based upon evidence and not just "gaps" in our understanding, they never presented clear criteria by which this evidence could be recognized. For example, no objective, scientifically useful definition of "irreducible complexity" was ever proposed by ID supporters. In arguing that complex adaptations could not have evolved gradually, ID advocates conveniently ignored hypotheses (which dated back to Darwin, who was very familiar with the argument) that features of organisms can be co-opted and change their function as evolution proceeds, and thus intermediate forms can be highly functional (see Chapter 5, Figure 9). ID depended on science's current inability to propose adequate Darwinian evolutionary explanations for every feature of every organism, and thus on the extent of scientific knowledge at the time, ignoring the obvious fact that scientific knowledge and understanding are continuously growing. Advocates of ID maintained that their view did not require a specific candidate (for example, God) in the role of designer, but a supernatural creator/designer of the sort they needed was clearly a god in all but name.

Modern ID was and is not science, but part of a larger cultural movement, the stated goal of which is to fundamentally alter the nature of science and society by removing materialism and making supernaturalism and religion legitimate parts of science. ID was a religiously motivated idea. This is clear from documents produced by ID advocates.

In 1999, ID had its first high-profile public victory in the U.S. when the Kansas Board of Education adopted statewide guidelines that removed references not only to evolution but also to the big bang, a universe billions of years old, and the geologic time scale. In 2000, however, voters removed the conservative members of the Board who had favored these changes, and state guidelines changed back to more evolution-friendly language. In 2005, however, with newly-elected conservatives once again in control, the Board reinstituted creationist-friendly changes, the most notable of which was a revision in the definition of science itself. The language in the standards was changed from "Science is the human activity of seeking natural explanations for what we observe in the world around us" to "Science is a systematic method of continuing investigation that uses observation, hypothesis testing, measurement, experimentation, logical argument and theory-building to lead to more adequate explanations for natural phenomena." Scientists across the country pointed out that the omission of the phrase "natural explanations" was significant because it would appear to open the way for non-natural explanations to be considered in the science classroom. On February 13, 2007, the Kansas State Board of Education rejected the 2005 revision, reestablishing science as restricted to the investigation of physical phenomena.

Also in 2005, a group of parents in Dover, Pennsylvania, sued their local school board because of an October 2004 Board policy that said, in part: "Students will be made aware of gaps/problems in Darwin's Theory and of other theories of evolution including, but not limited to, intelligent design." Plaintiffs argued that the policy was an unconstitutional breach of the wall between church and state. The suit, *Kitzmiller et al. v. Dover Area School District*, represented the first legal test of ID in the public school classroom. On December 21, 2005, in a sweeping opinion, Federal Judge John Jones III ruled for the plaintiffs that ID is religion and not science, and cannot be taught in the public schools. Judge Jones' 135-page opinion is a resounding condemnation of ID.

Modern creationism is thus quite diverse. It includes people who think the Earth is 10,000 years old, and those who believe it is much older; people who believe that the Biblical flood explains all of the geological record, and those who accept a more complex history. Although most people who consider themselves creationists reject evolution in any form, there are others who accept that life has evolved, but only under the direct guidance of God, as well as some who accept evolution by natural selection for all living things except humans. All creationists, however, believe in the action of divine or supernatural forces in shaping the natural world on a regular basis.

At this writing (January 2009), creationism is once again largely out of the public spotlight in the United States. The Dover decision effectively destroyed the credibility of ID, making it highly unlikely that any other community will adopt this approach to introducing creationism into public school classrooms. The results of the 2008 U.S. elections, furthermore, suggest that the country is undergoing yet another political sea change, this time in a more liberal, or at least moderate, direction, which historically has signaled that creationism retreats from center stage. Yet creationism is certainly not dead. As described in the following chapter, a majority of Americans still say

Figure 24. *Popular creationism. A banner (center) outside a church in upstate New York advertising a creationist exhibit that explains why dinosaurs became extinct (they didn't fit inside Noah's Ark), with a sample of recent books by creationist authors.*

they hold creationist beliefs. Creationist organizations and activists continue to try to cause problems for the teaching of evolution at local and state levels. An emerging approach is to call for requirements that students be exposed to the "weaknesses" of Darwinian evolution. Such requirements are often put forth under the cover of innocuous-sounding terms like "critical analysis," making it seem like all that is being requested is objective analysis of scientific alternatives.

Relatively little attention in the U.S. and Europe has been placed on non-Christian and non-Western resistance to and criticism of evolution. With more attention in the West being focused on Islam, however, it is becoming clear that many fundamentalist Muslims have as much trouble with evolution as many fundamentalists Christians. Turkey, for example, has the most active antievolution creationist movement of any country outside the United States.[12]

Why Scientists (and Others) Object to Creationism

Americans generally pride themselves on their fairness and tolerance for multiple points of view. It therefore strikes many nonscientists as curious and perhaps suspicious that scientists and others are so insistent that only evolution be taught in science classrooms, to the exclusion of creationist ideas. Why do scientists and their allies take this position? There are at least two reasons, both related to the fundamental methods and philosophy of science.

First, the various claims of creationism have repeatedly been tested scientifically – for hundreds of years – and they have failed. Thus to continue to teach it as a viable or potentially viable competing scientific hypothesis is to act as though all of this didn't happen. Science is not about considering all ideas all of the time. It is about testing ideas against observations, and discarding those that don't measure up as well. The central tenets of creationism – that the Earth is only a few thousand years old, that life has always been as we see it today, that a single worldwide flood was responsible for most or all of the geological record in the Earth's crust, that there are an insufficient number of "transitional forms" in the fossil record, and that an intelligent designer is required to explain complex adaptations – have been tested many times as scientific hypotheses in the past, and they have been falsified. There is no evidence that they are valid scientific ideas, and therefore they have no place in the science classroom. Creationism doesn't belong in the science classroom for the same reasons that astrology, alchemy, or the Earth-centered solar system are not: all of them have been tested and discarded and science has moved on in more productive directions.

Second, if evolution is wrong, so is a lot of other science. Creationists do not usually discuss this, preferring to claim that only evolution and/or natural selection are flawed. Yet if evolution (driven mostly but not wholly by natural selection), which is accepted as the dominant explanation for the order, history, and diversity of life by essentially every knowledgeable scientist in the world, is wrong, then there is likely something fundamentally misguided about most of the rest of science, from astrophysics to geology to molecular biology. Recent ID advocates have taken this issue even farther, claiming that one of their movement's "governing goals" is to "defeat scientific materialism and its destructive moral, cultural, and political legacies" and "to replace materialistic explanations with the theistic understanding that nature and human beings are created by God."[13] They do not, however, address what they would do about all of the materialistic science that they apparently do not object to, such as medicine, agriculture, and the various branches of physics and chemistry that make it possible for them to drive cars and use computers.

Like all attempts at "scientific" creationism, advocates of ID want it both ways: materialistic science when it suits them, and supernatural intervention when it doesn't, with no objective rules or regularities to explain or predict why one and not the other in any particular case.

Summary - Creationism:

- Modern creationism spans a broad range of ideas with respect to the occurrence of evolution and the age of the Earth, but has at its core the belief that one or more elements of the natural world were designed and/or made by a divine, supernatural force.

- Modern creationism is not science. It is a political/social movement.

- Creationist explanations for geological and biological phenomena have been repeatedly tested scientifically, over several centuries, and they have failed – they have been falsified – as scientific hypotheses. To claim otherwise is simply a lie. This doesn't mean that creationist ideas are wrong; it just means that they have not been validated by science, and therefore do not belong in science classrooms any more than other wholly discarded hypotheses.

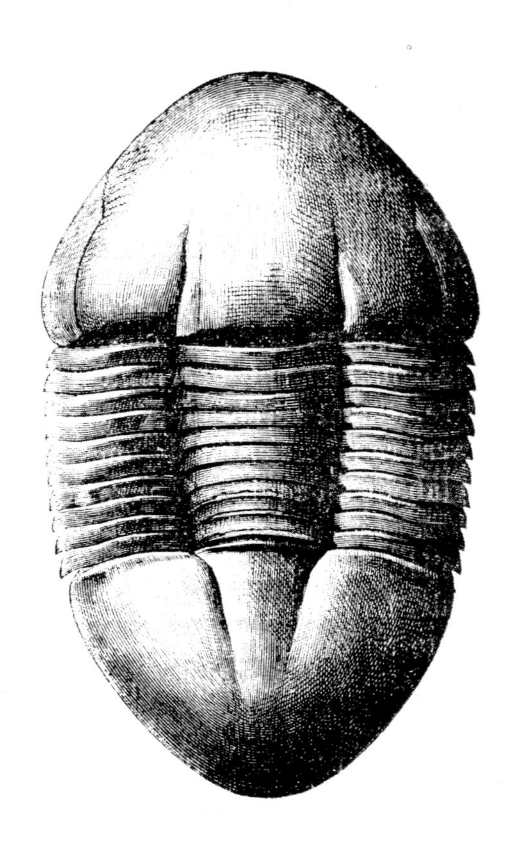

8. CURRENT AMERICAN PUBLIC VIEWS OF EVOLUTION

Although results of public opinion polls vary considerably depending on wording of individual questions (and the results are not always internally consistent), such polls have been generally consistent over more than 20 years in reporting that most people in the United States do not accept evolution, that is, they do not think that evolution happened or that evolution explains the natural world.

In a November 2004 poll by CBS, for example, 55% of those responding agreed with the statement "God created humans in present form," whereas 27% agreed that "Humans evolved; God guided the process," and 13% agreed that "Humans evolved; God did not guide the process." The same poll found that 65% of Americans thought that evolution and creationism should both be taught, whereas 37% thought that creationism should be taught instead of evolution. In a July 2005 Harris poll, 54% did not believe that human beings developed from earlier species (up from 46% in 1994), 49% believed that plants and animals have evolved from some other species (45% did not believe that), and 46% believed that apes and humans have a common ancestor (47% believed we do not).

Another 2005 poll reported that 42% of American adults believed that "life has existed in its present form since the beginning of time," whereas 48% believed that "life has evolved over time." Of these, 18% thought that this evolution was "guided by a supreme being," whereas 26% believed it was "evolution by natural selection." Only about half of the respondents to two 1999-2000 surveys answered "false" to the statement, "The earliest humans lived at the same time as the dinosaurs." Less than half responded "true" to the statement, "Human beings, as we know them today, developed from earlier species of animals."

On the other hand, a large majority in several recent polls said that they think evolution should be taught in public schools, and some polls suggested that, although a majority of Americans might prefer that creationism be taught, they do not want it taught as science. A 2001 poll by the National Science Foundation found that close to 80% of American adults agreed with the statement "the continents on which we live have been moving their location for millions of years," which would seem supportive of a general evolutionary viewpoint.

Some of the disagreement between these results is surely due to deep and wide public misunderstanding of evolution. Americans do not really know much about evolution. For example, a 1993 International Social Survey poll revealed that of 21 nations surveyed on people's basic knowledge of evolution, Americans were last, behind Bulgaria and Slovenia. In a 2004 Gallup poll, 30% said they didn't know enough about evolution to have any opinion on it. In a 2000 poll by People for the American Way (PFAW), 34% agreed with an incorrect definition of evolution ("humans have developed from apes over the past millions of years" - a more accurate statement would be that humans and apes share a common ancestor) and another 16% either thought evolution means something else or didn't know what it means. Americans similarly don't know much about what scientists think of evolution. A 2004 *Newsweek* survey, for example, found just 45% of respondents think that evolution is widely accepted by the scientific community and well supported by evidence. In the 2000 PFAW poll, of the 95% who had heard of evolution, 69% believe either that "you can never know for sure" or that evolution is a "mostly" or "completely not accurate account of how humans were created and developed."

A poll by CBS in late 2006 showed that: "Americans do not believe that humans evolved, and the vast majority says that even if they evolved, God guided the process. Just 13 percent say that God was not involved. ... Support for evolution is more heavily concentrated among those with more education and among those who attend religious services rarely or not at all." The poll found that 55% said that "God created humans in [their] present form; 27% said they believed that Humans evolved, [but] God guided the process," and 13% said they believed that "humans evolved [but] God did not guide [the] process."[14]

Summary - Current American Public Views:

- Essentially all national polls indicate that a majority of people in the United States do not accept evolution, that is, they do not think that evolution happened or that evolution explains the natural world.
- Polls also indicate that most Americans do not know much about evolution, and that much of what they think they know is incorrect.

Figure 25. *God creates Adam according to Michaelangelo, on the ceiling of the Sistine Chapel in The Vatican, Rome.*

9. EVOLUTION & RELIGION

Initial reactions to Darwin's arguments by most organized religions and clerics in Europe and North America were largely (although not exclusively) negative, but not in a particularly extreme or organized fashion.[15] Much more serious and organized religious criticism developed only in the early 20th century, largely in the U.S., coincident with a surge in the popularity and influence of fundamentalist Protestantism. From the 1920s through the 1960s, this resistance to evolution held sway without much challenge across much of the American religious landscape.

By the late 1960s, however, much in America had changed, and religious opinion on evolution and/or natural selection had become much more diverse. Many larger Christian denominations in Europe and the U.S. came to terms with evolution and took the official position that evolution and faith are compatible. These accommodations differ in details, but the basic approach has been along the lines of the statement attributed to Galileo that "science is about how the heavens go, whereas religion is about how to go to heaven." In other words, science is about the nature of the material world, whereas religion deals with the ethical, moral, and spiritual about which science can say nothing.

In 1969, for example, the United Presbyterian Church in the U.S.A. stated, and in 1982 and 2002 reaffirmed, its official position that "there is no contradiction between an evolutionary theory of human origins and the doctrine of God as Creator." In 1992, the United Church of Christ stated that "We acknowledge modern evolutionary theory as the best present-day scientific explanation of the existence of life on Earth; such a conviction is in no way at odds with our belief in a Creator God." Other major denominational organizations, including the American Jewish Congress, the Central Conference of American Rabbis, the General Convention of the Episcopal Church, the Unitarian-Universalist Association, and the United Methodist Church all have over the past 25 years issued statements opposing teaching "scientific creationism" in the public schools. The plaintiffs in the landmark 1981 case *McLean v. Arkansas*, who successfully sought to overturn the imposition of "balanced treatment" for creationism and evolution in the public school classroom, included Jewish, United Methodist, Episcopal, Roman Catholic, African Methodist Episcopal, Presbyterian, and Southern Baptist clergy.[16]

In 1996, Pope John Paul II proclaimed that the theory of Darwinian evolution is so well supported by so much evidence that it has become "more than just a hypothesis." Evolution, said the Pope, is fully

Figure 26. *The American botanist Asa Gray (1810-1888) accepted evolution and natural selection but urged Darwin to accept both. (Courtesy of the Cornell University Library.)*

compatible with Christian faith and a valid explanation of the development of life on Earth, with one major exception: the human soul. "If the human body has its origin in living material which preexists it," the Pope said, "the spiritual soul is immediately created by God."[17]

Since 1990, a number of scholars have presented views that attempt to accommodate theistic religion – the existence of a meaningful God – with acceptance of evolution. Advocates of these views, including many practicing scientists, historians, and theologians, believe that despite its apparent purposelessness, Darwinism does not imply a Godless universe. Natural selection, in this view, is simply how we can describe God's wider vision and wisdom and the "law-like behavior" of a "continuing creation."

A bit farther along the spectrum away from theistic religion are the views of many scientists and religious liberals, who believe that religion and evolution are compatible, but only if the religion does not require an active personal interventionist God. Religion, in this view, is the realm of human-defined purposes, meanings, values, and ethics. A particularly full statement of this view was given by paleontologist Stephen Jay Gould (1941-2002), who argued that the proper relationship of science and religion is "respectful noninterference ... between the two distinct subjects, each covering a central [yet distinct] facet of human existence."[18] Gould cited Darwin himself as a supporter of this view. Darwin kept his religious views confined to close family and friends. His private writings reveal that at the end of his life he was an agnostic, perhaps an atheist, who simultaneously held strong views on ethics, morals, and values. In 1860, six months after publication of the *Origin of Species*, Darwin wrote to the American botanist Asa Gray (1810-1888; Figure 26), who accepted evolution and natural selection but urged Darwin to accept both as instituted by God for His own purposes:

> *With respect to the theological view of the question. This is always very painful to me. I am bewildered. I had no intention to write atheistically. But I own that I cannot see as plainly as others do, and as I should wish to do, evidence of design and beneficence on all sides of us. There seems to me too much misery in the world ... On the other hand, I cannot anyhow be*

contented to view this wonderful universe, and especially the nature of man, and to conclude that everything is the result of brute force. I am inclined to look at everything as resulting from designed laws, with the details, whether good or bad, left to the working out of what we may call chance. Not that this notion at all satisfies me. I feel most deeply that the whole subject is too profound for the human intellect. A dog might as well speculate on the mind of Newton. Let each man hope and believe what he can.[19]

Several denominations, especially fundamentalist Protestant sects in the U.S., have definitely not made peace with evolution, and some of these are at the center of modern opposition to evolution, for example, in public schools. Outside of Europe and the U.S., as mentioned above (Chapter 7, page 73) many fundamentalist Islamic and Hindu sects and clerics are similarly hostile to evolution and/or natural selection. Followers of these faiths believe that the acceptance of Darwinian evolution requires the total triumph of materialism and the rejection of any theistic religion – that is, any belief system that involves an intelligent designer, a point or purpose to the universe, or a larger meaning to existence. Interestingly, this is the opinion of not just many religious creationists, but also many leading evolutionary biologists as well.

So is evolution compatible with religion or not? Opinion is divided. On the one hand, some religious traditions find the materialistic/naturalistic assumptions that underlie Darwinism and all other science morally unacceptable, and many evolutionary biologists similarly hold that it is impossible to accept both a Darwinian view of the evolutionary process and any seriously theistic religious view. On the other hand, there are many evolutionary biologists who profess to be somewhat or very religious, and there are numerous clergy, denominations, and religious individuals who see no conflict between their faith and evolutionary science. It is clearly possible to hold the view (as many practicing scientists do) that science and religion need not be in conflict with each other, because they address fundamentally different aspects of human experience. Science deals only with material reality. Religion deals with the spiritual, the moral, and the ethical. Many scientists profess that science cannot ever answer ultimate questions such as "why are we here," "what was the beginning of everything," or "how should we live our lives." According to this view, these questions very properly belong in the realm of religion.

Summary - Evolution & Religion:

- The relationship between evolution and religion is more complex than is usually presented in the media and popular opinion.
- Many Christian and Jewish denominations and organizations, including The Vatican, have publicly stated that there is no conflict between belief in God and acceptance of evolution by natural selection.
- Many scientists, including some evolutionary biologists and paleontologists, say that they accept both Darwinism and a personally meaningful God, demonstrating that it is possible to do so. The philosophical status of such statements, however, remains a subject of controversy.

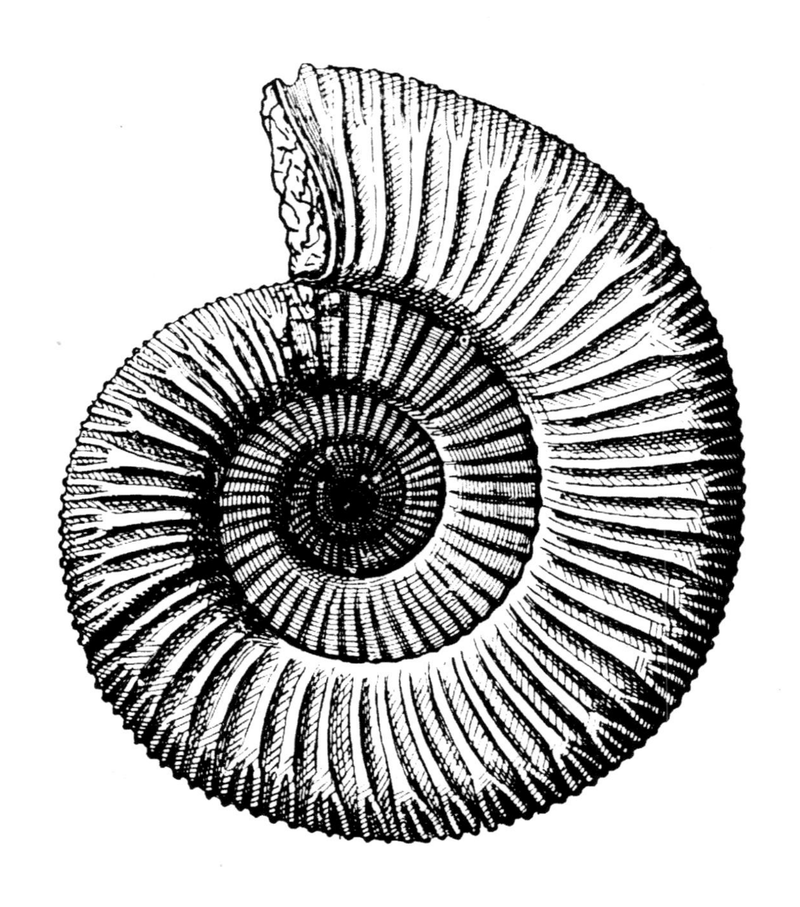

10. WHY DOES EVOLUTION MATTER?

Whatever one thinks of Charles Darwin and the ideas he expounded, it is indisputable that few individuals in history have had so much impact on humanity's view of itself, its world, and its place in that world. Ever since publication of the *Origin of Species* in 1859, it has been clear that humans can never again see the world in quite the same way as they had before. This recognition started even in Darwin's lifetime. It continued over the following century with Darwin's routine inclusion in virtually every list of the most influential thinkers in history, and the *Origin*'s in virtually every list of the five or ten or 20 most important and influential books ever written. Darwin's name appears on scores of imposing public buildings in many countries, alongside those of Aristotle, Plato, Copernicus, Newton, and Einstein, testimony to the fact that it is instantly recognizable to millions of people as among the highest aristocracy of science and knowledge. Around the world, the words "evolution" and "natural selection" – and images of Darwin and evolution – are ubiquitous in literature, advertising, and popular culture.

What is the source of this fame? It is not just that Darwin reorganized biology. The broader impacts of Darwin's two arguments (descent with modification and natural selection) lie both within and outside of science, and this is true even among people who say they do not understand or accept what Darwin said. In many important respects, for more than a century the world (at least the Western world) has been in large part a Darwinian world: we all live within a worldview that is thoroughly imbued – scientifically and culturally – with ideas first laid out convincingly by Charles Darwin in 1859. Yet we are so close to this fact that we usually do not generally recognize it. As writer Gillian Beer has put it, "We pay Darwin the homage of our assumptions."[20]

Darwin and Science

The explanatory power and implications of evolution and natural selection for basic scientific questions about living things have been discussed in the preceding chapters. It is also important to point out, however, that Darwin not only proposed the first serious explanations for life's order, history, and diversity. He also revolutionized biology as a whole, and set it on the path that would eventually lead it to the hugely important role it plays today.

When Darwin was an undergraduate student at Cambridge University in the 1820s, it was not possible to take what we would today call "science" courses for credit. Biology outside of medicine was part of "natural philosophy," and was almost completely the avocation of wealthy gentlemen (the word "scientist" was not even invented until the 1840s). Even medicine was hardly what we would call "scientific." (Charles Darwin's father was a successful physician, and told his son that there was often actually very little that he could do to help his patients other than listen to them.) By the end of the 19th century, in contrast, biology was an established and professionalized branch of science, and medicine had been transformed into a field far more dependent on the results of biological science.

Although he was not solely responsible for the late 19th century rise of biology as a successful separate branch of science, Darwin was unquestionably a crucial part of this ascent. Even though natural selection was not immediately embraced in detail, its form as an explanation and the style of its presentation made it suddenly clear that biology could be just as rigorously empirical and scientific as physics and chemistry. Natural selection was simple and plausible, and, most importantly, it was completely materialistic and nonteleological – it did not require supernatural intervention. And Darwin presented it with extraordinary attention to logic and empirical detail, a form of argument heretofore restricted to the physical sciences. "It is hardly an exaggeration," wrote philosophers Vittorio Hösle and Christian Illies, "to state that Charles Darwin has shaped biology more significantly than anyone else before or since. His importance can be compared only with that of Aristotle."[21] Darwin was biology's Newton.

Essentially every modern field of biology concerned with whole organisms (as opposed to their parts, such as cells or chemicals), including ecology, behavior, and systematics (the study of biodiversity) is based on evolution. Virtually all practicing professional biologists who work on whole organisms accept evolution as an adequate explanation for the order, history, and diversity of life they observe. There is no serious disagreement among such professional biologists about whether evolution is "true." Even biologists who do not work on whole organisms, such as molecular biologists, cell biologists, and medical researchers, must use evolution if they pursue questions about why the structures and functions that they are studying are as they are.

More specifically, evolution is especially relevant to several areas of applied biology. Almost all of the food you eat was shaped by evolution – under the influence of human-mediated artificial selection in a manner exactly analogous to natural selection. Agriculture *is* evolution. The antibiotic your doctor prescribes for your child's ear infection is contributing to selection on the bacteria in his or her body and in the environment as a whole, which is why many bacteria are becoming resistant to many antibiotics. The same applies to pesticides and insects that damage crops. The plants and animals we have introduced into parts of the world where they did not naturally occur – think of starlings, zebra mussels, or kudzu in North America, rabbits in Australia, giant African snails in Hawaii, or rats and cats in New Zealand – have literally altered the evolution of countless other organisms in these regions. The current acceleration of human-caused environmental change – from deforestation to climate change – is altering the physical environment for every other species on Earth, and leading to the premature extinction of many. Humans have become a force of change of geological proportions. We are causing evolution in ways that we can only vaguely predict.

The 21st century has been predicted to be the "century of biology," when knowledge of the life sciences will play an inescapably large role in economic and social life, and in decisions of enormous environmental and economic significance. An understanding of evolution, in particular, will play a crucial and increasing role in medicine, agriculture, and conservation.[22] It is difficult to imagine that the United States can remain competitive in such times with half of its population not accepting the central organizing principle of biology. More generally, watering down or eliminating evolution will lead inevitably to the diminution of all science, with potentially catastrophic consequences for economic growth.

Evolution is relevant to far more than biology. Evolution is also central to many areas of the Earth sciences, such as stratigraphy (the study of the layering of rocks), geochronology (geological dating), tectonics (the study of mountain building and other major Earth movements), and paleontology (the study of the history of life as revealed by fossils). Furthermore, the assumptions and conclusions from other areas of science that underlie evolution – such as the great age of the universe, the solar system, and the Earth, the continuity of past and present processes, and the constancy of physical law in time and space – are shared with other fields of science such as astronomy, physics, and chemistry. This is why every major organization of professional scientists in the United States has endorsed the teaching of evolution.

Thus within science, what Darwin did is clear: he very quickly convinced his fellow scientists that evolution was true, and laid the groundwork for the eventual acceptance that natural selection was its primary cause. By so doing, he made the scientific study of whole organisms possible. Outside of science, however, Darwin's influence was and is perhaps even greater, although more complicated to understand.

Exploration of Darwin's extrascientific impact follows two broad lines of scholarship. One is the explication and documentation of the actual historical influences of evolution and natural selection on fields such as literature, religion, philosophy, the social sciences, and general perceptions of history and humans' place in nature. The other is the rapidly growing number of calls for new and more rigorous applications of natural selection to a wide variety of other areas, such as medicine, psychology, economics, linguistics, political science, and international relations. These two are not equivalent. The first addresses what has already happened, whereas the other argues for influences or applications that have only begun to affect thought and practice in any major way, or which could, in the future, be seen to have become important influences.

Historical Influences

Darwin was clearly both a reflection and a cause of larger changes in society and culture. His "scientific arguments were," as Rutgers English professor George Levine put it, "part of a whole movement of which Darwin can be taken as the most powerful codifier."[23] We can clearly recognize and group the influence of these arguments under at least six headings.[24]

(1) The rise of science. Whether we like it or not, the dominance of science (and its technological offspring) is one of the central defining features of modern life. Darwin was both follower and leader of the currents that created this situation. Today, when we largely take for granted – or at least consider the strong possibility – that most, if not all, subjects can be treated by the methods of science, it is perhaps difficult to imagine a time or a world view in which this was not so. Darwin obviously did not single-handedly put science in the central role it plays today, but, as Levine put it, Darwin "can be taken as the figure through whom the full implications of the developing authority of scientific thought began to be felt by modern nonscientific culture." Darwin's passionate friend and defender Thomas Henry Huxley concluded his 1860 essay on the *Origin* by predicting that Darwin's ideas would exert a large influence, "not only on the future of Biology, but in extending the domination of Science over regions of thought into which she has, as yet, hardly penetrated."[25] And so they have.

(2) Human nature and place in nature. In the midst of the rapid advance of science and technology that marked Darwin's time, human beings were largely exempt. Medicine was (very slowly) becoming more mechanistic and scientific, but virtually every other aspect of humans was off-limits to science. Darwin changed this forever, and made humans a legitimate scientific subject. He proposed the first serious scientific theory that explained the origin, history, and nature of humans – what was at the time often referred to as "man's place in nature," and he laid out the techniques by which these topics could be studied empirically. Although the first edition of the *Origin* had only a single mention of humans ("Light will be thrown on the origin of man and his history,"[26]), it was clear to virtually every reader what the larger implications of Darwinian evolution were for humans. Not everyone agreed with these implications, of course, but Darwin made them acceptable topics of serious scientific discussion.

Darwin's influence marks the beginning of much of what today are called the social sciences, including psychology, sociology, and economics. We take for granted that these fields can address issues of human thought and behavior from an objective, observational, and secular (*i.e.*, "scientific") point of view. This is not all Darwin's doing. Others before him had considered what we would now call "social evolution" of various sorts. However, the creation of these modern fields of inquiry would not have happened without him. "The authority of science," wrote Levine, "and its extension from natural phenomena to human was both a condition of Darwin's enterprise, and its consequence."[27]

The implications of evolution and natural selection for philosophy were both immediately apparent and very subtle, and their impact was both sudden and still emerging today, a century and a half later. Darwin's antiteleology contributed to the decline of idealism and romanticism, which were still widespread in mid-19th century philosophy in both Europe and America. Darwin's idea of a continuous "struggle for existence" implied that values were not inherent or absolute, and that nothing could persist if it could not maintain itself in its environment. Darwin's emphasis on empirical experience argued for a new approach to seeking continuity between humans and the natural world, which many previous philosophers had found only in the world of thought.

The imminent practicality of Darwin's scientific approach strongly affected the thought of a number of influential philosophers and thinkers, particularly in America, including especially William James, Charles Sanders Pierce, Oliver Wendell Holmes, and John Dewey. These "pragmatists" argued for the importance of ideas that were workable over those that were more ethereal and absolute.[28] The entire field of ethics was essentially thrown open (a condition in which it largely remains). No longer was it self-evident that ethical standards could come only from revealed religion. After Darwin, wrote Levine, "Value would now be seen to inhere not in permanence, but in change, not in mechanical design but in flexibility and randomness ... Once the consonance between the natural and the intentional is lost, the space for willed constructions of meanings ... opens up."[29] Humans, in other words, must seek – and make – meaning for themselves as best they can.

(3) Religion. When nonscientists think of Darwin and his influence, they frequently think of God and religion (see Chapter 9). Although it is true that many religious people, from 1860 to the present, have argued vehemently that evolution and/or Darwinian natural selection are utterly irreconcilable with belief in any meaningful God or adherence to any traditional religious faith, this is not by any means true for all religious people or traditions. Despite suggestions to the contrary, Darwin obviously did not destroy religion. Indeed, persistent predictions about the imminent disappearance of religion have all proven utterly incorrect. It is also true, however, that, at least in most Western countries, organized religion today has less direct impact on the daily lives of people, governments, and institutions across the industrialized world than it did a century and a half ago. Although clearly not all of this change was caused by science in general, or Darwin in particular, both were clearly involved. As conservative as we now think Victorian culture was, the period was in fact a time of pervasive secularizing of nature and society and in the exploration of the consequences of that secularization. As Levine noted:

> *The tradition of natural theology was threatened and largely dismantled by Darwinian science, and in the process nature, society, narrative, and language itself were desacralized, severed from the inherent significance, value, and meaning of a divinely created and designed world. The Darwinian quest for origins was the signal and the authoritatively scientific means by which fact was severed from meaning and value ... and the world had to be reconstituted not from divine inheritance but from arbitrary acts of human will.*[30]

(4) Literature. Although literary criticism is often caricatured as the epitome of socially irrelevant, academic navel-gazing, the nature of a culture's fiction literature is also taken as one of the clearest windows into its core values, tastes, and assumptions. If this is true, then the impact of Darwin – as revealed in his influence on at least American and British literature – is great indeed. Just in the past quarter-century, literary scholars have produced detailed demonstrations of Darwinian influences on the writing of many authors, including Joseph Conrad, Charles Dickens, George Eliot, William Faulkner, Robert Frost, Thomas Hardy, D. H. Lawrence, Bram Stoker, and H. G. Wells.[31] These influences mostly consisted of subtle yet pervasive structuring of plot and themes, such as humans' interactions with each other or with nature. Their ubiquity is further evidence that Darwin was only a part, albeit a large one, of a larger current of social and intellectual change. Novelist and historian Henry Adams devoted an entire chapter of his famous 1907 autobiography to "Darwinism."[32] Although Adams had a complex and sometimes conflicting set of responses to Darwinism, he claimed (probably partially tongue-in-cheek) that he was "a Darwinian for fun," suggesting that, whatever one might think of it, Darwinism was by the early 20th century an integral part of intellectual culture.

(5) Change and chance. Scholars continue to debate whether Darwinism was more of a product or a cause of the undeniable shift toward a more secular, scientific, and constantly changing society that occurred in the West in the late 19th and early 20th centuries. Whatever its cause, there is no doubt that this shift occurred and that Darwinian evolution was part of it. From language and literature to philosophy and politics, in both the natural and social sciences, the concept of ubiquitous and ceaseless change – driven largely or exclusively by natural processes – became increasingly accepted after 1859 and is one of the most obvious characteristics of modern Western culture and society. "Obviously," wrote Levine, "the theme of change did not need Darwin to invent it. But in his world everything is always or potentially changing, and nothing can be understood without its history." Darwin's world "is in constant process. Adaptation is but for the moment." Darwin, as Levine put it, in effect fundamentally changed "the way his culture could think."[33] We now live in this culture of constant change.

(6) Dystopia. "It was easy," wrote Levine with considerable understatement, "to use Darwinism to serve a multiplicity of antithetical purposes." Darwin's emphasis on imperfection as evidence of evolution (see discussion of vestiges in Chapter 4) was a major departure from the tradition of harmony emphasized by natural theology. Instead of all having been created for some higher, greater good, Darwin, as Levine said, saw "adaptation as contingent and incomplete, however breathtakingly wonderful it can be. He demonstrates disharmony, maladaptation, imperfection." This was, perhaps understandably, seen by some as a prescription for a harsh and grim view of the present and future of humanity. In a world in which only the strong survive, then the dominance of the powerful over the weak was not just advantageous for those on the top of the socio-economic ladder, it was the proper and natural state of the world. This kind of thinking was part of the basis not only for the so-called "social Darwinism" of social scientists such as Herbert Spencer and industrialists like Andrew Carnegie, but also for the racist and imperialist policies of the late 19th century British Empire and the mid-20th century Nazi regime, as well as the popularity of eugenics in the United States during the 1920s.[34]

Such applications of Darwin's ideas are now virtually unanimously condemned by scientists and politicians alike, but they are frequently cited even today by critics of evolution as evidence of its negative impact on society. Darwin himself largely rejected such extrapolations. Coming as he did from a liberal background, Darwin was personally on the progressive end of the sociopolitical spectrum of his time. Although he did not doubt the "superiority" of white Europeans over various other racial groups, he did not think that this conclusion was an appropriate basis for social policy.[35]

The question of whether the use of science and technology – whether evolution or nuclear energy – for destructive or "evil" purposes means that science itself is a potentially negative force is a persistent one for philosophers and politicians alike. Yet the fact that arguments made by Darwin can be used by people to justify horrific acts has absolutely no bearing on whether those arguments can help us to better understand the nature and history of life, which they clearly do. It does mean, however, that we should all be knowledgeable and vigilant enough to be able to question the application of all scientific ideas to areas beyond those for which they were originally formulated. This brings us to the second major line of thought about the impact and influence of Charles Darwin.

Universal Darwinism?

Although Darwin's influence has extended, as just described, across the width and breadth of human thought over the past 150 years, there were many fields and areas of research that considered, and then backed away from, a strict application of natural selection. For example, in 1960 the noted historian of evolution John Greene wrote that if Darwin could "view the contemporary intellectual scene" he would find in the social sciences "evolutionary problems largely neglected and his own theory of social progress through natural selection in great disfavor. In the current emphasis on man's uniqueness as a culture-transmitting animal," Greene continued, Darwin might even "sense a tendency to return to the pre-evolutionary idea of an absolute distinction between man and other animals."[36]

Greene's assessment no longer holds. The past quarter-century has seen a dramatic expansion of the application of natural selection to virtually every area of human activity, including those that had de-emphasized it as an adequate explanatory approach just a few years before. Much of this new fondness for natural selection as an essentially "universal" explanation for all features of living things – human and otherwise – can be traced to the explication in the late 1970s of the field of "sociobiology" by entomologist and evolutionary biologist Edward O. Wilson, and it has been particularly conspicuous in psychology and the social sciences, but has also extended to the arts and well beyond.[37] Unlike the historical influences described above, however, this influence of Darwin is still in many respects novel and controversial. It is the subject of a large and growing literature, and I will here only touch on a few of its most notable purported accomplishments.

(1) Psychology. Although not anti-Darwinian, the schools of thought that dominated psychology during most of the 20th century (Freudianism, Jungianism, etc.) emphasized nongenetic, "environmental" factors as much as or more than inherited factors on which natural selection might act. In contrast, modern "evolutionary psychology" attempts to explain virtually all human (and nonhuman) mental phenomena – from aggression and love to religion and infidelity – as the results of (mostly past) natural selection. The mind is the way it is, argue advocates of this view, because natural selection built it that way.[38]

(2) Literature. According to a growing school of literary critics, not only are some aspects of the novels of Conrad or the poems of Frost influenced by evolutionary themes, virtually all human relationships described in literature can be analyzed and understood as results of natural selection. The jealousy of Othello is ultimately the result of the differing reproductive demands on males and females and of biologically mandated competition for mates. The hubris of Macbeth and emotional turmoil of Hamlet are similarly the results of genetically-determined predispositions – albeit more or less affected by their various environmental contexts – for particular behaviors.[39]

(3) International relations and national security. In these times of international turmoil, terrorism, and uncertainty, it is perhaps not surprising that the new Darwinism has turned its attention to natural selection as a potential explanation for patterns of relationships among peoples and nations. In some cases, such analyses are little more than simple analogies between societal security problems and characteristics evolved in nonhuman nature, which are used to shed light on strengths or weaknesses of particular societal security arrangements or systems. Other analyses focus on how actual evolutionary processes, such as the development of the ancestral mind and the emergence of human social structures, might be affecting the security environments today. Still others attempt to use tools that were developed for evolutionary and ecological studies – such as demographic and epidemiological models – to address security problems. In such analyses, war and peace, just to take one example, are explained as results of natural selection for the survival of individuals or groups, in addition to, or instead of, the results of purely non-genetic, "social" factors.[40]

Medicine

Finally, mention should be made of a major area that appears to be no longer in the "maybe" column of having been powerfully influenced by Darwinian thought: medicine. Since the 1980s, a field variously known as "evolutionary" or "Darwinian medicine," built around the notion that disease can be understood and treated from an explicitly Darwinian (*i.e.*, evolution driven mainly by natural selection) point of view – has grown from an exotic suggestion to the mainstream. Darwinian medicine addresses topics such as antibiotic resistance of bacteria and viruses, inherited or partly-inherited conditions such as obesity and high blood pressure, and diseases such as cancer and mental illness. It has its own robust research literature and is now taught at several major medical schools.[41] (Ironically, surveys of American physicians suggest that a much larger proportion do not fully accept Darwinian evolution than in any other biological discipline. A 2005 poll of 1472 physicians, for example, found that, whereas 78% of them accept evolution, 63% of Protestant doctors believe that intelligent design is a "legitimate scientific speculation."[42] It remains to be seen what effect the increasing status of evolutionary medicine will have on such views.)

Summary - Why Does Evolution Matter?:

- Evolution matters historically because it played a key role in several major changes in Western culture and society, including revolutionizing the entire field of biology, altering how humanity views itself and its relation to the world, a general increase in secularity, and significant changes to the arts and social sciences.

- Evolution matters today because it provides crucial understanding for major applied areas of biology such as agriculture, conservation, genomics, and medicine.

- Evolution is a central assumption and tool of many scientific fields outside of biology, including geology and astronomy.

- Evolution by natural selection has exerted profound influence on nonscientific fields from literature and philosophy to economics and art. Application of Darwinism to additional fields, including psychology and international relations, is currently extremely active, ensuring that evolution and natural selection will continue to exert profound influence in the wider culture into the foreseeable future.

May 17, 2001
Mr. Harnyk
Sample
57
48
Mr. Harnyk
(Paleontologists)
Journal Entries
Lunch
SSR
Outside
Study Time
Dismissal
16

11. Teaching Evolution

Pre-college teachers, especially in American public schools, occupy an important position in the process of improving the basic understanding of evolution. Public schools, again especially in the United States, are also often the flash points for controversy about evolution, and they are the source for most of the popular media coverage on the topic. As discussed in the preceding chapter, evolution is about much more than biology. It is therefore important for teachers in all subjects to be familiar with it and with some of the most common issues that arise in its teaching. Parents, school board members, administrators, and elected officials should also be informed enough about the subject to be able to participate intelligently in discussions when they occur.

Evolution in American Public Schools

Residents of other countries are often mystified by American public education, not least because, instead of a single mandated national curriculum, such as exists in many countries, the U.S. has 50 separate sets of standards, one for each state. Within each state, furthermore, local school boards can have considerable influence on what is taught. This decentralized structure means that discussions over the teaching of evolution in American public schools can (and do) take place in literally thousands of separate official venues around the country, and virtually guarantees that it will never be settled with any finality for the nation as a whole.

Although, as described in Chapter 7, the teaching of evolution in American public schools has increased and improved considerably since the 1960s, it is still very inconsistent among, and even within, the states. For example, in a recent evaluation of state science standards that took evolution explicitly into account, 19 states (including over half of all U.S. children) received overall grades for their science standards of A or B. Fifteen states, however, received failing grades. The remainder received grades of C or D. For evolution in particular, 20 states were judged to have "sound" scores in their treatment (signifying that evolution was required, encouraged, and supported in their official curricula and standards), down slightly from 24 in 2000. The number of states earning "passing" grades in evolution remained at 7, whereas those earning "marginal" grades rose from 6 to 10. Failing grades in evolution (signifying that there was little or no mandated treatment of evolution at all) stayed steady at 13.[43]

In public school classrooms, the situation looks even grimmer. Although there are thousands of teachers who do an exemplary job of putting evolution at the center of their teaching about biology and Earth science, there are clearly many thousands of teachers who do not. Studies in the 1980s, for example, suggested that many school board members, and almost half of the science teachers in the United States, support including at least some creationism in the curriculum, and that perhaps as many as half of the nation's public high school students were receiving educations shaped by significant creationist influence. More recent work indicates that the situation has not changed significantly. For example, a 2000 poll reported that 79% of educators supported teaching creationism in schools. In the late 1990s, only 57% of science teachers nationwide considered evolution to be "a unifying theme in biology," and almost half of all science teachers said that they believe that there is as much evidence for creationism as there is for evolution. In many states, biology teachers often place little or no evidence on evolution. Still more recent reports in the media indicate that these trends are continuing.[44]

Better training for teachers produces, at best, uneven results. In a recent study of 44 pre-certified secondary biology teachers before and after they completed a 14-week graduate-level course on evolution, the researchers reported that although the course "produced statistically significant gains in teacher knowledge of evolution and the nature of science and a significant decrease in misconceptions about evolution and natural selection," the majority of them "still preferred that antievolutionary ideas be taught in school."[45] Numerous previous studies support this result.

Nor does college training seem to help. Even among undergraduates who take one or more courses in which evolution is covered, there appears to be a disconnect between knowledge and belief. In a 1993 study, for example, a group of 1,200 college students was tested at the beginning of the semester on their evolutionary knowledge. Biology majors scored only 6% higher than nonmajors in their knowledge. When the same students were retested on the first day of the following semester, after the biology majors had taken a semester-long course on evolution, the researchers were surprised to find that "majors, who received a much more rigorous treatment of the material, came through the semester with the same degree of understanding as the nonmajors!"[46]

The challenges facing these generally poorly prepared teachers are substantial. The teaching of evolution is under considerable negative pressure, even in states in which standards are excellent. For example, studies have found that up to 20% of teachers report that they have been pressured to not teach evolution.[47] Students frequently challenge teachers on various aspects of evolution, often coming armed with creationist-supplied information or internet documents such as "Ten Questions to Ask your Biology Teacher about Evolution." Others express their opposition to evolution on homework assignments. Many just tune out. The stakes are extremely high. As a 2008 *New York Times* article on teaching evolution in Florida put it, how science teachers confront these challenges "could bear on whether a new generation of Americans embraces scientific evidence alongside religious belief."[48]

Some Suggestions for Teachers

(1) Prepare but don't ignite. Arrange your curriculum or lesson plans to anticipate and address potential common misconceptions about evolution, but do not assume that all or even most students necessarily know enough to have misconceptions. In other words, do not start off defensively; be positive.

(2) Use what you have available, and the comparative method. You don't need perfect fossils or DNA sequencers to teach evolution. All you need is any material that shows a feature of an organism – a feather, a fossil shell, a bone, a leaf, a beetle, or a butterfly – and a little bit of information about its natural history. What do related or similar forms look like? Where do they live? Is there a known fossil record for the group? Have relatives been domesticated? Comparison of just two different organisms might be all you need to set up the basic structure of questions about where features and species come from.

(3) Get back to basics. The use of the word *theory* in everyday language is very different than its use in scientific realms. To most, a theory refers to a hunch, or best guess. Many people therefore dismiss evolution as "just a theory" that has little evidence to support it. However, in science – all sciences – the term *theory* refers to an idea that has stood up to rigorous testing by scientists the world over. A theory is an explanation that is well supported by the evidence, and those related to evolution are no different. In any science curriculum, from elementary to college, it is always beneficial to review the basic terms and language of science.

(4) Clearly distinguish between occurrence and mechanism of evolution (*i.e.*, descent with modification versus natural selection). Many seemingly interminable controversies around the teaching of evolution are evidence of the continuing widespread confusion between the conclusion that evolution (meaning "descent with modification") has occurred and is responsible for the patterns we see in living thing, and the mechanisms or causes of evolution. It has been stated repeatedly earlier in this book, but clearly needs repeating, that there has been *no* serious debate among knowledgeable biologists or geologists about whether evolution occurs since Charles Darwin died in 1882. This doesn't mean it's true, just that essentially every serious scientist working in the field accepts that it is. There continues, however, to be vigorous debate over the causes of evolution, and in this sense (as one contributor put it)"Nothing is evolving faster than evolutionary theory." Darwin's theory of natural selection continues to be overwhelmingly accepted by scientists as the dominant cause of evolution, but many other mechanisms are also clearly involved. "Teach the controversy" is a valid approach to evolution education only if there is a controversy. And there simply is none in this case.

(5) Emphasize that evolution explains observations and answers questions. Darwin did not come to the conclusion that species had changed through time because he particularly wanted to, or because he had some larger agenda. He accepted evolution because it answered satisfactorily so many pressing questions about life. This is why evolution is accepted by virtually all knowledgeable scientists today: because it works. If another scientific hypothesis that does a better job at answering these questions comes along, evolution could be discarded. However, all lines of available evidence – from comparative anatomy to embryology to genetics – support evolution as *the* explanation for life's diversity on Earth, past and present.

(6) Don't equivocate. Evolution isn't just an idea that "some scientists" have, or just "one view" among many. It is *the* theory of biology that is essentially universally accepted by all scientists that know anything about the subject. Natural selection as a mechanism for evolution is almost as widely accepted. To tell students otherwise – for example, that "scientists aren't sure" or that there are "weaknesses" in the theory – is simply lying to them. This isn't because we know absolutely that evolution by natural selection is true – it's because there is so much evidence for it that it would be ridiculous not to accept it.

(7) Know the material. This sounds obvious, but it is especially important in the case of evolution because teachers might be less frequently or conspicuously challenged by students on other subjects. Saying "I don't know" if you don't is always the best approach, but whether students are honestly seeking answers or trying to play "gotcha," hesitation or nervousness on the part of the teacher when students ask questions can signal exactly what some of the students suspect – that the evidence for evolution really is shaky and alternatives are equally plausible. Being able to articulate quick definitions and distinctions, such as are described earlier in this book for Darwin's two arguments or categories of evidence, can often be extremely useful for impressing students that you know what you're talking about.

(8) Seek help when (or, ideally, before) you need it. Like all scientific fields, evolutionary biology is a large and complex affair. Just learning the basics can be challenging enough, much less keeping up with new developments. However, there are abundant resources for teachers who need more information. Some of the best of these are listed in the end of this book. Teachers might also feel that they need another kind of help – when they are challenged by parents or school boards about their teaching of evolution. There is help available here too. The National Center for Science Education is a not-for-profit organization that exists to assist teachers and others dealing with the challenges of creationism. NCSE is also an excellent source of up-to-date information on evolution-creationist controversies. Contact information is also provided in the back of this book.

(9) Review your own beliefs. Students are often interested in a teacher's personal or religious views on evolution. Many teachers, however, particularly if they have been trained as scientists and do not come from strong religious traditions, might not have articulated their own philosophical perspectives. Many who do come from religious backgrounds might never have questioned the specifics of their faith. It is a good idea to do so, even if you never actually communicate these views to your students, because it can stimulate you to clarify your presentation of how and why evolutionary science works. Do you believe God exists? Why or why not? If you do believe in God, what role do you believe God plays in structuring the natural world? Answering these for yourself – even writing them out – is an important step in being able to help your students in their own explorations and struggles.

Summary - Teaching Evolution:

- Teaching evolution honestly and accurately is one of the most important jobs in modern education. Evolution is not an educational frill or option. It is one of the cornerstones of scientific literacy.

- Many pre-college biology and Earth science teachers are either woefully unprepared or actually resistant to teaching evolution, and it is not clear how this situation can be improved in the near future.

- Willing teachers can improve their teaching of evolution in a variety of ways, including focusing on a few central messages, practicing quick definitions and speaking without apology so as not to suggest inappropriate uncertainty, and clarifying – if only for themselves – their own personal beliefs about the relationship between religion and evolutionary science.

- There is help available for teachers who want it in the form of numerous books and websites, as well as advice and support from the National Center for Science Education (www.ncseweb.org).

12. SOME FREQUENTLY ASKED QUESTIONS

What is evolution?

Organic evolution is the idea that all organisms are connected by genealogy and have changed through time.

How does evolution happen?

Evolution is probably caused by several processes, the most important of which is natural selection.

What is "Darwinism?"

Darwinism is sometimes used as a synonym for "evolution" or "descent with modification," but this is incorrect. Darwinism refers to evolution mainly caused by natural selection, as described by Charles Darwin in *On the Origin of Species* in 1859. Evolution can be the result of causes other than natural selection. Scientists are highly confident that evolution has occurred, and that Darwinian natural selection is perhaps the most important mechanism of the evolutionary process as we currently understand it. Scientists continue to investigate other mechanisms relating to evolutionary change.

Is evolution "just a theory?"

A "theory" in science is a structure of related ideas that explains one or more natural phenomena and that is supported by observations from the natural world. It is not something less than a "fact." Theories actually occupy the highest, not the lowest, rank of certainty among scientific ideas. They are systems of explanation that unite many different kinds of data and observations. Theories can be modified when new information becomes available, and they can be overturned or discarded when evidence to the contrary becomes so overwhelming that it can no longer be explained away. Evolution is a theory in the same way that the idea that matter is made of atoms is a theory, that bacteria cause disease is a theory, that the sun being the center of the solar system is a theory. Any of these theories might be incorrect (and good scientists must always consider that possibility), but scientists accept all of them as provisionally true because there is so much evidence to support them.

Is Darwinian evolution "random?"

No. Darwin's favored cause of evolution, natural selection, is highly directional, and not random at all. Natural selection means that this directionality is provided by the environment, which "selects" variants that do better at surviving and reproducing. The underlying genetic variation, according to this theory, is random only in the sense that it is not in any preferred direction relative to the direction of eventual evolutionary change. Variation, in Darwin's view, is in all directions, and then the environment "steers" it down only one or a few routes. Furthermore, natural selection is cumulative – it does not start from scratch every generation. Thus statements such as "the chances of assembling a human being by chance are astronomical" are irrelevant – change by natural selection happens incrementally, generation by generation. No evolutionary biologist has ever argued that "a human being has been assembled by chance."

Is it true that there is lots of evidence against evolution?

No. Essentially all available data and observations from the natural world support the hypothesis of evolution. No credible biologist or geologist today doubts whether evolution occurred. Debate continues, however, among scientists about the causes of evolutionary change.

Aren't there lots of scientists who don't accept evolution?

Although there are indeed serious scholars who do not accept evolution, few of these are practicing biologists or geologists who are actually doing real research and seeking honestly to understand how aspects of the Earth and its life came to be as they are. Various lists of scientists who "doubt" or "dissent" from evolution notwithstanding, there are tens of thousands of professional practicing scientists who effectively test evolution every day and as a result believe it is the best available explanation for the history, order, and diversity of life.

What is the difference between microevolution and macroevolution?

Microevolution is evolutionary change within a species, such as we observe in domesticated plants and animals, pesticide and antibiotic resistance, or laboratory experiments. It is also the kind of evolutionary change observable in the wild, such as beak-shape changes in birds subjected to harsh climate changes. Macroevolution is evolution among many species and across long series of ancestor-to-descendant connections, which scientists believe takes place over thousands to millions of years.

How can evolution work if most mutations are harmful?

Typically genetic mutations are harmful, but these are for the most part quickly purged from populations by natural selection. Many genetic mutations cause little or no change. Some mutations, however, are beneficial, as shown by numerous experiments in the lab and in nature. The complex adaptations that so impress us, however, are not the result of single mutations but are based on combinations of mutations that have been shown (experimentally and in the wild) to increase in frequency because the individuals that carry them incur some advantage in life that results in their leaving more offspring over the course of their lifetimes than individuals without the mutations.

How do we know how old all rocks and fossils are?

Strictly speaking, the age of the Earth (or its rocks or fossils) in years isn't really relevant to whether or how evolution occurred. As long as the Earth is very old (more than a few hundred million years), evolution by natural selection could have occurred. However, the process by which geologists determine the age of rocks and fossils is very important because fossils provide some of the best evidence for evolution, and the best record of the history of life on Earth. Geologists assign relative dates to rocks and the fossils they contain by using superposition and correlation. They usually assign numerical dates to rocks by analyzing radioactive elements inside mineral grains in volcanic rocks, using a process called radiometric dating. There are many kinds of radiometric dating, used for rocks of different ages and often providing independent tests. Other complementary methods of obtaining numerical ages also exist, which give consistent data with radiometric dating.

Doesn't the complexity/design of nature imply an intelligent designer?

Science deals only with material causes of material phenomena. Intelligent design is not testable scientifically, and nothing we can observe in nature requires a supernatural designer. We therefore defer to observable, measurable processes to explain the patterns we see in nature.

Is evolution against religion?

Not necessarily. The most often-cited evidence for this is the fact that there are many evolutionary biologists and paleontologists who profess to be somewhat or very religious. More generally, it is possible to hold the view (as many practicing scientists do) that science and religion need not be in conflict with each other, because they address fundamentally different aspects of human experience. Science deals only with material reality. Religion deals with the spiritual, the moral, and the ethical. Many scientists profess that science cannot ever answer ultimate questions such as "why are we here," "what was the beginning of everything," or "how should we live our lives." According to this view, these questions very properly belong in the realm of religion.

What's wrong with teaching both evolution and creationism in the science classroom?

The central tenets of most kinds of creationism – that the Earth is only a few thousand years old, that life has always been as we see it today, or has been created from nothing numerous times, that a single world-wide flood was responsible for most or all of the geological record in the Earth's crust, that there are unbridgeable gaps in the fossil record and between different modern kinds of living things – have been tested as scientific hypotheses in the past, and have completely failed. There is no evidence that they are valid scientific ideas, and therefore they have no place in the science classroom. These ideas continue to be promoted not because of their scientific validity, but because of their religious significance. To teach creationism in whatever form as science, based on the evidence that now exists, would be similar to teaching alchemy in chemistry class, or blood letting in medical school. It would seriously damage the training of students in science, putting the future of society at risk.

13. 10 Things Everyone Should Know About Evolution

(1) Evolution (descent with modification) is not the same as mechanisms or causes of evolution.

(2) Scientists can study events and processes in the past, even when there was no human there to witness them, by the principle of extrapolation, which is also used in all scientific experiments and conclusions about phenomena that happen in the present.

(3) There is a huge amount of evidence for evolution (descent with modification) from all areas of biology, and there has been no serious scientific debate about whether it is true since the 1880s. Evolution is as well supported as many other scientific conclusions we regularly call "facts," such as that the Earth goes around the Sun.

(4) Very active scientific debate continues today about the mechanisms by which evolution occurs, but this does not imply any controversy about whether evolution occurs.

(5) Since the 1940s, natural selection, the mechanism first proposed by Charles Darwin in 1859, has been accepted by most scientists as the most important mechanism of evolution.

(6) Natural selection is simply an outcome of the struggle of variable individual organisms to survive and reproduce. Those individuals with inherited variations that allow them to be more successful in this struggle will, on average, leave more offspring in the next generation.

(7) Natural selection does not guarantee progress or improvement in any absolute sense, nor does it include or imply any overarching plan. It only leads to better fit of organisms to their local environments.

(8) Evolution is the central organizing and explanatory idea in modern biology, including medicine and agriculture. If it is incorrect, then so is much of our understanding of the rest of biology.

(9) Evolution is important because it helps us understand and address many practical problems, such as resistance of germs and pests to medicines and pesticides, the structure and function of the genetic mechanisms of inheritance and development, the nature of many human diseases, and the ecological causes and consequences of species extinction.

(10) Evolution by natural selection is not necessarily opposed to religion, nor is it a basis for rejecting all systems of ethics. It does, however, imply that the natural world, including humans, is explicable solely by reference to natural processes and phenomena, that any supernatural influence on nature is unobservable by and inaccessible to science, and that human ethics and values are derived from humans themselves.

SOURCES OF MORE INFORMATION

Books & Articles

Alters, Brian J., & Sandra M. Alters. 2003. *Defending Evolution: a Guide to the Creation/Evolution Controversy*. Jones and Bartlett, Boston, 261 pp. [An extremely useful and authoritative source for educators, especially at the middle school and high school levels.]

Bowler, Peter J. 2003. *Evolution: The History of an Idea, 3rd ed.* The University of California Press, Berkeley, 483 pp. [An excellent and readable history of evolutionary thought, with an excellent bibliography.]

Coyne, J. A. 2009. *Why Evolution is True*. Viking Adult, New York, 320 pp. [A well-known evolutionary geneticist at the University of Chicago concisely summarizes the evidence for evolution, and shows why creationism is not supported by the evidence and also falls outside the bounds of science.]

Dawkins, Richard. 1996. *The Blind Watchmaker: Why the Evidence of Evolution Reveals a Universe Without Design*. W. W. Norton, New York, 400 pp. [The least polemical and most original of Dawkins' many books, in which he eloquently makes the case for design without a designer.]

Forrest, Barbara, & P. R. Gross. 2004. *Creationism's Trojan Horse: The Wedge of Intelligent Design*. Oxford University Press, New York, 401 pp. [A detailed but very readable analysis of the arguments of intelligent design and why they fail. The senior author provided key testimony in the 2005 Dover ID trial.]

Futuyma, Douglas J. 2005. *Evolution.* Sinauer Associates, Sunderland, Massachusetts, 543 pp. [An updated and streamlined version of the leading college-level textbook on evolution.]

Gould, Stephen J., ed. 2001. *The Book of Life: an Illustrated History of the Evolution of Life on Earth, 2nd ed.* W. W. Norton, New York, 256 pp. [An excellent introduction to how paleontologists think about evolution and the history of life, with great illustrations.]

Isaaks, M. 2005. *The Counter-Creationism Handbook.* Greenwood Press, Westport, Connecticut, 330 pp. [A very useful compilation of responses to 400 of the most common creationist claims.]

Jones, John, III. 2005. *Kitzmiller et al. v. Dover Area School District.* Memorandum Opinion (December 20, 2005). U.S. District Court for the Middle District of Pennsylvania, http://www.sciohost.org/ncse/kvd/kitzmiller_decision_20051220.pdf. [Judge John Jones' opinion in the Dover intelligent design trial missed no opportunity to explain that ID is not science and has no place in the science classroom. Highly recommended.]

Lewin, Roger. 1982. *Thread of Life: The Smithsonian Looks at Evolution.* W. W. Norton, New York. [An excellent semipopular overview of evolution over the 3.5 billion year history of life, emphasizing macroevolution and with excellent illustrations.]

National Academy of Sciences. 1998. *Teaching about Evolution and the Nature of Science.* National Academy Press, Washington, D.C. [The official statement of the U.S. National Academy of Sciences on the teaching of evolution.]

National Academy of Sciences. 1999. *Science and Creationism: a View from the National Academy of Sciences, 2nd ed.* National Academy Press, Washington, D.C. [The official statement on science and creationism and why they are not the same.]

Numbers, Ronald L. 2006. *The Creationists. From Scientific Creationism to Intelligent Design, expanded ed.* Harvard University Press, Cambridge, Massachusetts, 606 pp. [A detailed history and analysis of modern creationism by a noted historian of science.]

Pallen, Mark. 2009. *The Rough Guide to Evolution.* Rough Guides, London, 346 pp. [A thorough, original, and authoritative entry into the increasingly crowded "short guide" niche of evolution books, with an especially good treatment of the wider impact of evolution on modern culture and society.]

Pennock, Robert T. 1999. *Tower of Babel: The Evidence Against the New Creationism.* Massachusetts Institute of Technology Press, Cambridge, Massachusetts, 429 pp.

[A thorough and detailed critique of modern "intelligent design" theory by a philosopher.]

Petto, A. J., & L. R. Godfrey, eds. 2007. *Scientists Confront Intelligent Design and Creationism.* W. W. Norton, New York, 463 pp. [A collection of very readable essays by noted scientists on a wide range of topics around evolution and creationism.]

Pigliucci, Massimo. 2002. *Denying Evolution: Creationism, Scientism, and the Nature of Science.* Sinauer Associates, Sunderland, Massachusetts, 338 pp. [An excellent and accessible summary of the subject by an author with Ph.D. degrees in both biology and philosophy, including especially good discussions of why creationist arguments are not correct.]

Prothero, Donald R. 2007. *Evolution: What the Fossils Say and Why It Matters.* Columbia University Press, New York, 381 pp. [A very useful and accessible compilation of examples from the fossil record that support the idea of evolution, written by a well-known vertebrate paleontologist.]

Schneiderman, Jill S., & Warren D. Allmon, eds. 2009. *For the Rock Record: Geologists on Intelligent Design.* University of California Press, Berkeley, 264 pp. [A group of geologists and paleontologists provide perspectives on the problems of intelligent design and the implications for understanding and teaching evolution, paleontology, and geology.]

Scott, Eugenie. 2004. *Evolution vs. Creationism: an Introduction.* Greenwood Press, Westport, Connecticut, 272 pp. [The Executive Director of the National Center for Science Education gives a clear overview of the arguments for and against evolution, the approaches used by modern creationists, and selections from the primary literature.]

Zimmer, Carl. 2001. *Evolution: The Triumph of an Idea.* Harper Collins, New York, 364 pp. [The companion volume to the PBS television series.]

Books for Younger Readers

Eldredge, Niles, G. Eldredge, & D. Eldredge. 1989. *The Fossil Factory: a Kid's Guide To Digging Up Dinosaurs, Exploring Evolution, and Finding Fossils.* Roberts Rinehart Publishers, Lanham, Maryland, 111 pp. [Written by famous paleontologist Niles Eldredge and his two teenage sons, this is a kid-friendly exploration of paleontology, geology, and evolution, complete with activities to perform at home.]

Gamlin, L. 2000. *Evolution.* Dorling Kindersley, New York, 64 pp. [Patt of the *Eyewitness* series that explains evolution and the history of life.]

Jenkins, S. 2002. *Life on Earth: The Story of Evolution.* Houghton Mifflin, New York, 40 pp. [Jenkens gears this picture book of the history and diversity of life toward children aged 6 to 10.]

Lawson, K. 2003. *Darwin and Evolution for Kids: His Life and Ideas, with 21 Activities.* Chicago Review Press, Chicago, 146 pp. [Part biography, part natural history, part activity book for children in grades 5 through 9.]

Peters, L.W. 2003. *Our Family Tree: an Evolution Story.* Harcourt, San Diego, California, 48 pp. [This picture book geared toward young children tells the history of life on Earth from single-celled organisms to humans.]

Stein, S. 1986. *The Evolution Book.* Workman Publishing, New York, 389 pp. [This book tackles the four-billon-year history of life for older students, grades 6 through 10.]

Strauss, R. 2004. *Tree Of Life: The Incredible Biodiversity of Life on Earth.* Kids Can Press, Tonawanda, New York, 40 pp. [Strauss describes biodiversity with the concept of a family tree of life and includes a section on habitat protection and one for parents, teachers, and guardians.]

Webster, S. 2000. *The Kingfisher Book of Evolution.* Kingfisher, New York, 96 pp. [Geared toward children in grades 5 through 9, this book tackles subjects such as the evolution of behavior, humans, and the future of evolution, as well as research on cloning, gene therapy, and extraterrestrial evolution.]

Websites

The National Center for Science Education – http://www.ncseweb.org
The NCSE is the leading clearing house and advocacy organization that monitors creationist activity around the country and provides information on evolution and creationism for educators at all levels. Their website is full of timely information and links to additional on-line resources.

Understanding Evolution (University of California Museum of Paleontology) – http://evolution.berkeley.edu
A "one-stop shop" for information on evolution. Highly recommended.

Paleontological Research Institution and its Museum of the Earth – http://www.museumoftheearth.org
PRI is a natural history museum and research institution in Ithaca, New York. Its mission is to increase and disseminate knowledge about the history and evolution of life on Earth. It is home to one of the largest fossil collections in the United States, which is utilized by professional researchers and students in many fields. PRI's website contains information about the collection, evolution and creationism, and its exhibit facility, the Museum of the Earth, which is open to the public.

GLOSSARY

adaptation – a feature of an organism that assists it in survival or reproduction and that resulted from natural selection for its current function; also the process by which such features are built by natural selection.

biogeography – the study of the geographic distribution of organisms.

biological succession – in geology, the observation that different fossils occur in different layers of sedimentary rocks; also referred to as "geological succession."

comparative anatomy – the study of similarities and differences in the body parts of different species of organisms.

constraint – any condition of genetics or development that limits or restricts the production or expression of heritable variation that is available to natural selection.

co-opting – a change in the function of a characteristic of an organism that can lead to the evolutionary development of a new characteristic.

correlation – in geology, proposal of the hypothesis that two layers of sedimentary rock are the same age, usually based on their similar fossil content.

creationism – the belief that the Earth and its life were created by a supernatural power.

descent with modification – change of organisms over time within an ancestor-descendant connection; Darwin's term for what is today generally referred to as "evolution."

diversity – the variety of life forms on Earth, usually the number of species; often called "biodiversity."

DNA – deoxyribonucleic acid, the complex organic molecule that is the genetic material of almost all living things on Earth.

embryology – the study of embryos, the earliest stages in the life-history of organisms, and the changes that they undergo as they grow into adults.

evolution – the theory that organisms are connected by genealogy and have changed through time.

evo-devo – evolutionary developmental biology, a relatively new sub-discipline of evolutionary biology that studies the connections between the changes that a single individual organism goes through during its life (from embryo to adulthood) and the changes that species or other groups go through over millions of years.

extinction – complete disappearance of all individuals in a population, species, or larger related group of organisms.

extrapolation – the process of inferring or projecting from what is known to what is not known.

fact – in science, a statement or theory supported by so much evidence that it would be perverse to withhold provisional consent.

fitness – in evolutionary genetics, the rate of increase of its descendants in later generations.

fossil – the remains or trace of an organism that lived in the geological past preserved in the crust of the Earth.

gene – a unit of the material of inheritance; a section of DNA that contains instructions for some function of a living cell and that is inherited by descendant cells and organisms.

genetic drift – the accumulation of random changes in a gene pool.

genetics – the science of heredity or inheritance.

Geological Time Scale – the internationally-agreed-upon series of names that refer to periods of time in the geological past.

hypothesis – an idea proposed to explain a natural phenomenon.

inheritance – in genetics, the passing along of characteristics from ancestor to descendant by transmission of genetic material.

Intelligent Design – the idea that features of the physical universe and/or life can be best explained by reference to an "intelligent cause" rather than a natural process or material mechanism.

macroevolution – evolution among or above the level of species, happening over thousands to millions of years.

materialism – an approach to understanding that uses only physical observations and physical causes to account for phenomena; also called "naturalism."

microevolution – evolution within species, happening over as little as a few generations up to hundreds or thousands of years.

Modern Synthesis – the comprehensive theory of evolution developed in the 1940s by collaboration between geneticists, systematists, and paleontologists; also called "Neodarwinism."

mutation – abrupt change in the DNA or RNA of an organism.

natural selection – a cause of evolution in which one or more consistent differences in fitness (*i.e.*, survival and reproduction) among groups of organisms leads to changes in their genetic characteristics.

numerical dating – any of a number of techniques in geology for determining the age of minerals, rocks, or other geological phenomena in years before present.

overproduction – in biology, the observation that more individuals are commonly born than will survive to adulthood.

paleontology – the study of the history of life using fossils.

population genetics – the mathematical study of gene frequencies and their changes within populations of organisms.

provisional – in science, a term applied to a temporarily accepted idea, based on information currently available, and that could be rejected if sufficient contrary information becomes available.

punctuated equilibrium – a theory of evolution that suggests that species of organisms arise relatively suddenly, and then show little or no significant change during their duration.

radioactive decay – the process by which atoms of certain chemical elements spontaneously break down, emitting radiation and changing into atoms of other chemical elements.

radiometric dating – any of a number of techniques in geology that make use of the clocklike regularity of radioactive decay of elements in minerals to provide numerical estimates of the age of geological phenomena.

relative dating – the establishment of the ordering in time of two or more geological phenomena.

RNA – ribonucleic acid, a complex organic molecule used in most organisms to transmit information from the genes (DNA) to other parts of the cell, especially in the manufacture of proteins.

science – an approach to explaining and understanding the natural world that seeks physical causes for physical phenomena by testing hypotheses using observations of matter and energy.

speciation – the evolutionary process by which one species splits into two or more separate branches, giving rise to at least one additional species.

species – a group of organisms with its own coherence and evolutionary history; among sexually reproducing animals, a group that reproduces among themselves but not with others.

species sorting – a theory of macroevolutionary cause in which long-term trends result from differences in the rates of origination and extinction of species, rather than transformations within species.

stasis – in paleontology, the observation that a species shows relatively little change in form during its duration.

struggle for existence – Darwin's term for the necessity of individual organisms striving to survive and reproduce; caused by overproduction.

superposition – in geology, the principle that in a stack of undisturbed layered sedimentary rocks, the oldest layers are on the bottom and overlying layers are sequentially younger.

systematics – the field of biology concerned with naming and classifying organisms; the study of the diversity of life.

teleology – the supposition that there is purpose or directive principle in the works and processes of nature.

test – a method for evaluating a scientific hypothesis that compares predictions to observations about the physical world.

theory – an idea or set of ideas that connects, explains, and is supported by a large number of observations.

transformation – evolutionary change of an entire population or species, in contrast to its division into separate groups that thereafter have separate evolutionary trajectories (*i.e.*, speciation); evolutionary biologists commonly refer to this as "anagenesis" and speciation as "cladogenesis."

trend – in evolution, a pattern of net change in form over a long period of geological time.

variation – observed differences among individual organisms in a population of a species.

vestige – a feature of an organism that does not make sense in terms of current function and is a similarity shared with other different organisms.

ENDNOTES

[1] Before 1859, the concept of organisms changing through time was usually referred to as "transmutation." Darwin's phrase "descent with modification" importantly added a genealogical connection. Evolution, in other words, is not just change – it is change within an unbroken series of ancestors and descendants. Darwin did not use the word "evolution" in the first edition of the *Origin of Species* (1859), although the last word of that edition is "evolved." Before Darwin, "evolution" referred to the process of embryological development, that is, change within rather than between generations. The word assumed its modern meaning after 1859, and Darwin adopted it in later editions of the *Origin*.

[2] Paley, William, 1805, *Natural Theology*, London, 595 pp; pp. 1-3.

[3] The "radical" nature of natural selection was discussed by Gould, S. J., 1977, *Ever Since Darwin*, W. W. Norton, New York, 285 pp.; and Dennett, D., 1996, *Darwin's Dangerous Idea: Evolution and the Meaning of Life*, Simon & Schuster, New York, 586 pp.

[4] The word "adaptation" has at least two meanings in evolutionary biology. In both cases, it refers to a feature of an organism that enhances its survival or reproduction. Some evolutionary biologists use the term for any such feature, no matter what caused it, whereas others emphasize the historical perspective, and restrict the term to such features that resulted from natural selection for a specific function. Still others would restrict the term further to only features built by natural selection for their current function.

[5] All of these categories except genetics correspond to chapters in Darwin's *Origin of Species*, which is still in many ways the best available compendium of evidence for descent with modification.

[6] Critics of evolution often say that they cannot accept it because no one has ever "turned a cat into a dog" or even "created a new species." Yet many of our domesticated animals and plants – which we know were descended from common ancestors – are so different from each other that if we were to encounter them in nature, we would surely call them distinct species if not higher taxonomic levels. Just think of how different Chihuahuas are from Great Danes, or broccoli from cauliflower, yet we know that these forms share a common ancestor.

[7] Darwin, Charles R., 1859, *On the Origin of Species by Means of Natural Selection, or the Preservation of Favoured Races in the Struggle for Life [1st ed.]*, Charles Murray,

London, 502 pp.; pp. 80-81.

[8] This term was and still is frequently misunderstood as implying a necessarily brutal physical competition between organisms, but Darwin intended by the word "struggle" only the idea that every organism must do its utmost to survive and reproduce.

[9] Darwin, Charles R., 1875, *The Variation of Animals and Plants Under Domestication, 2nd ed.,* John Murray, London, 2 volumes; vol. 1, pp. 6-7.

[10] Confusingly, the same pattern is commonly referred to as "biological succession" when speaking to geologists, and as "geological succession" when speaking to biologists.

[11] For an excellent account of the Scopes trial, see Larson, E., 1997, *Summer for the Gods: The Scopes Trial and America's Continuing Debate Over Science and Religion.* Basic Books, New York, 336 pp.

[12] See Edis, T., 2007, *An Illustion of Harmony: Science and Religion in Islam*, Prometheus Books, Amherst, New York, 265 pp.; Pallen, Mark, 2009, *The Rough Guide to Evolution*, Rough Guides, London, 346 pp.; p. 298.

[13] Discovery Institute, 2003, "The Wedge document": So what?, http://www.discovery.org/scripts/viewDB/filesDB download.php?id=349. (Document is dated February 3, 2006, but indicates that it was "originally published in 2003.")

[14] Robinson, B. A., 2008, "Public beliefs about evolution and creation," Ontario Consultants on Religious Tolerance, http://www.religioustolerance.org/ev_publi.htm.

[15] Some clergy in Darwin's time actually argued actively in favor of evolution. See, *e.g.*, Livingstone, D. N., 1984, *Darwin's Forgotten Defenders: The Encounter Between Evangelical Theology and Evolutionary Thought*, Regent College Publishing, Vancouver, British Columbia, Canada, 210 pp.

[16] Matsumura, M., ed., 1996, *Voices for Evolution, 2nd ed*, National Center for Science Education, Berkeley, California, 176 pp.

[17] Pope John Paul II, 22 October 1996, Message to the Pontifical Academy of Sciences on evolution, http://www.ewtn.com/library/papaldoc/jp961022.htm.

[18] Gould, S. J., 1999, *Rocks of Ages: Science and Religion in the Fullness of Life*, Ballantine, New York, 241 pp.; p. 5.

[19] Letter from Charles Darwin to Asa Gray, 22 May 1860, in: Burkhardt, Frederick, *et al.*, eds., 1993, *The Correspondence of Charles Darwin, vol. 8,* Cambridge University Press, Cambridge [see also http://www.darwinproject.ac.uk]; p. 223.

[20] Beer, G., 1983, *Darwin's Plots: Evolutionary Narrative in Darwin, George Eliot and Nineteenth-Century Fiction*, Routledge & Kegan Paul, London, 303 pp.; p. 5.

[21] Hösle, V., & C. Illies, 2005, *Darwinism and Philosophy*, University of Notre Dame Press, Notre Dame, Indiana, 392 pp.; pp. 1-2.

[22] See, *e.g.*, Mindell, D. P., 2007, *The Evolving World: Evolution in Everyday Life*, Harvard University Press, Cambridge, Massachusetts, 352 pp.

[23] Levine, G., 1988, *Darwin and the Novelists: Patterns of Science in Victorian Fiction*, Harvard University Press, Cambridge, Massachusetts, 319 pp.; p. 9.

[24] Pallen, *op. cit.*, p. 77 ff..

[25] Huxley, T. H., *Darwiniana Essays*, chapter 2, The Origin of Species (1860), pp. 22-79; p. 79 (quoted by Levine, *op. cit.*, p. 88).

[26] Darwin, 1859, *op. cit.*, p. 488.

[27] Levine, *op. cit.*, p. 14-15.
[28] See, *e.g.*, Menand, L., 2001, *The Metaphysical Club: a Story of Ideas in America*, Farrar, Straus and Giroux, New York, 384 pp.
[29] Levine, op. cit., p. 94.
[30] Levine, op. cit., p. viii.
[31] See, *e.g.*, Beer, *op. cit.*; Levine, *op. cit.*; Faggen, R., 1997, *Robert Frost and the Challenge of Darwin*, University of Michigan Press, Ann Arbor, 363 pp.; Wainwright, M., 2008, *Darwin and Faulkner's Novels: Evolution and Southern Fiction*, Palgrave Macmillan, New York, 243 pp.
[32] Adams, H., 1906, *The Education of Henry Adams*, Furst and Company. Reprinted in 2002 by Courier Dover Publications, New York, 384 pp. Text also available online at Project Gutenberg, http://www.gutenberg.org/etext/2044.
[33] Levine, *op. cit.*, p. 85.
[34] On "social Darwinism," see Hofstadter, R., 1959, *Social Darwinism in American Thought*, George Braziller, New York, 248 pp.; Bannister, R. C., 1979, *Social Darwinism: Science and Myth in Anglo-American Social Thought*, Temple University Press, Philadelphia, 292 pp. On eugenics, see Kevles, D., 1998, *In the Name of Eugenics: Genetics and the Uses of Human Heredity*, Harvard University Press, Cambridge, Massachusetts, 448 pp.; Black, E., 2007, *War Against the Weak: Eugenics and America's Campaign to Create a Master Race*, Dialog Press, New York, 550 pp.
[35] Desmond, A., & J. Moore, 2009, *Darwin's Sacred Cause: How a Hatred of Slavery Shaped Darwin's Views on Human Evolution*, Houghton Mifflin Harcourt, New York, 448 pp.
[36] Greene, John C., 1961, *Darwin and the Modern World View*, Louisiana State University Press, Baton Rouge, 141 pp.; p. 129-130.
[37] See, *e.g.*, Wilson, E. O., 1975, *Sociobiology: The New Synthesis*, Harvard University Press, Cambridge, Massachusetts, 697 pp.; Wilson, E. O., *On Human Nature*, Harvard University Press, Cambridge, Massachusetts, 260 pp.
[38] See, *e.g.*, Buss, D., 2007, *Evolutionary Psychology: The New Science of the Mind, 3rd ed.*, Allyn & Bacon, Boston, 496 pp; for an alternate view, see Rose, H., 2000, *Alas, Poor Darwin: Arguments Against Evolutionary Psychology*, Harmony Books, New York, 352 pp.
[39] Barash, D., & N. Barash, 2005, *Madame Bovary's Ovaries: a Darwinian Look at Literature*, Delacorte Press, New York, 272 pp.
[40] See, *e.g.*, Thayer, B. A., 2004, *Darwin and International Relations: On the Evolutionary Origins of War and Ethnic Conflict*, University of Kentucky Press, Lexington, 425 pp.
[41] See, *e.g.*, Nesse, R. M., & G. C. Williams, 1995, *Why We Get Sick: The New Science of Darwinian Medicine*, Crown, New York, 291 pp.; Trevathan, W. R., E. O. Smith, & J. J. McKenna, eds., 1999, *Evolutionary Medicine*, Oxford University Press, New York, 480 pp.; Stearns, S. C., & J. C. Koella, eds., 2008, *Evolution in Health and Disease, 2nd ed.*, Oxford University Press, New York, 374 pp.
[42] Louis Finkelstein Institute for Religious and Social Studies at the Louis Stein Center (2005).
[43] See Allmon, W. D., R. Kissel, R. Ross, S. Sands, & T. Smrecak, 2009, "Teaching evolution in America: a status report on Darwin's 200th birthday, *American Paleontologist*, 17(1), and references therein.

[44] Allmon *et al.*, *op. cit.*

[45] Nehm, R. H., & I. Schonfeld, 2007, "Does increasing biology teacher knowledge about evolution and the nature of science lead to greater advocacy for teaching evolution in schools?" *Journal of Science Teacher Education*, 18(5): 699-723.

[46] Allmon *et al.*, *op. cit.*

[47] Allmon *et al.*, *op. cit.*

[48] Allmon *et al.*, *op. cit.*

Warren Allmon is Director of the Paleontological Research Institution in Ithaca, New York, and Hunter R. Rawlings III Professor of Paleontology in the Department of Earth and Atmospheric Sciences at Cornell University.